HOPE SPRINGS ETERNAL

Mead Botanical Garden is the legacy garden of the famous American horticulturist Theodore Luqueer Mead (1852 – 1936)

Garden Founders
Jack Connery and Edwin Grover

A History of
MEAD BOTANICAL GARDEN

Hope Springs Eternal

PAUL BUTLER

HOPE SPRINGS ETERNAL: A HISTORY OF MEAD BOTANICAL GARDEN

Copyright © 2019 by Paul Butler

Published by Little Red Hen Press, Winter Park, FL

All rights reserved. This book may not be reproduced in whole or in part, in any form (beyond copying permitted by Sections 107 and 108 of the United States Copyright Law, and except limited excerpts by reviewer for the public press), without written permission from Paul Butler. For permission, or to contact Paul Butler, please email casajard@gmail.com.

For a complete list of photo credits, please see *Picture Credits* at the end of the book.

Author services by Pedernales Publishing, LLC.
www.pedernalespublishing.com

Cover design: Paul Butler and Jose Ramirez

Library of Congress Control Number: 2019909383

ISBN 978-0-9979666-3-3 Hardcover Edition
ISBN 978-0-9979666-4-0 Paperback Edition

Printed in the United States of America

Dedication

This book is dedicated to the hundreds of committed volunteers who for more than 50 years have selflessly given their time, money and effort, striving to make Mead Botanical Garden the thing of beauty that it once was.

Epigraph

"To plant a garden is to believe in tomorrow"

Audrey Hepburn
British Actress
1929 - 1993

Contents

Preface	xi
Prologue – The Opening of the Garden	1
Chapter One – How the Garden Came About	7
Chapter Two – The Garden Develops Beautifully	53
Chapter Three – The Orchids of Mead Botanical Garden	81
Chapter Four – Financial Difficulties and the City Takes Over	99
Chapter Five – The Amphitheater and Fashions in the Garden	117
Chapter Six – The Roller-Coaster Years	129
Chapter Seven – Neglect and Destruction	151
Chapter Eight – A Flickering Renaissance	173
Acknowledgments	189
Notes	193
Picture Credits	233

Preface

This is the book I said I'd never write. Having completed a six-year writing odyssey culminating in the 2016 publishing of "Orchids and Butterflies," the biography of Theodore Mead, I was happy to leave the story of his legacy garden to the pen of others. Mead's story had flowed naturally from the collection of over 5,000 of his letters in the Winter Park Rollins College archives from which it was possible to discover just about everything that Mead was doing or thinking, almost daily. I believed that to write a book about a thing, rather than a person, was always going to be more difficult, particularly in the absence of original research material.

However, it began to dawn on me that if I didn't write it, no-one else would. Much of which had been written about the Garden's history was just plain wrong, and the role that John Connery, who preferred to be known as Jack, played in bringing the Garden about was rarely mentioned. The trigger came when I saw and read a 1995 letter from Jack's wife, Helen, to the Winter Park library archivist there at the time, Donna Rhein, in response to a request for information about Jack Connery and his involvement in the Garden.

In the letter, Helen writes, "I had felt that someday someone would be puzzled by all the misconceptions and misinformation – and lack of knowledge – floating around concerning the Mead Garden.… Some members of my family have urged me to write the history of the Garden, but it was a bittersweet period, and I have felt that I did not want to relive those years." Unfortunately Helen died in 2000, and with Jack having passed away in 1982, the opportunity for a first-person historical account was lost.

Edwin Grover and Jack Connery were the parents of Mead Botanical Garden. Both were unassuming characters that, although opposites, worked together well as a team. Grover was the intellectual, the writer, and the passionate and energetic promoter; Connery the organized and hard-working doer, with a strong desire to memorialize his mentor, Theodore Mead, as an outstanding American horticulturist.

Of the two founders, more is known about Grover and his achievements at Rollins College than Connery, although there is precious little insight into Grover the man, as acknowledged in his recent biography by Ed Gfeller. Judging by the few photographs we have of Grover, he was not a man given to lively expressions, neither for the camera nor in his writing. Nor could you imagine him using industrial language, but throughout the 1950s as the mismanagement of the Garden continued, he must have been close on occasions to using a few choice words from its vocabulary. At the start of writing this book, even less documentation could be found about the life of Jack Connery, who left the Garden in the early 1950s when the City of Winter Park took it over. All that was certain was that both he and his wife Helen were tireless workers in the Garden. They gave everything they had to try to make it successful to the extent that Helen, through stress and money worries, drove herself to exhaustion and eventual hospitalization with malnutrition.

The gaps in their story were fortuitously filled by a contact with Jack Connery's youngest son, Edwin, now 82 and living in Cape Coral, Florida. Edwin, his wife Nancy and daughter Cindy, shared with me facts and original research material offering an appreciation of intriguing aspects of Jack Connery's life and his role in the early stages of the Garden's development, as well as allowing the Connery's later life in DeLand to be accurately chronicled.

When it came to research material for the entire history of the Garden, there were two important primary sources – past copies of the Orlando Sentinel group of newspapers, and the historical records and scrapbooks of the Winter Park Garden Club, now housed in the archives of Rollins College library. The first source came out of the close relationship that Edwin Grover developed with Martin Andersen, owner and proprietor of the Sentinel for many years. The relationship ensured that Grover-written copy concerning the Garden and submitted to the newspapers inevitably got published, providing a rich source of information. For the second, the Winter Park Garden Club is to be commended on the detailed records and photographs they collected and meticulously documented over the crucial early years of the Garden's development.

The role of the City of Winter Park in the story of Mead Botanical Garden was a critical one and one that ended up not been able to be told in any detail. The devastation and destruction that they wrought upon the Garden in the 1970s, when most of the botanical material, all the greenhouses, and all the valuable Mead orchid collection suddenly disappeared, seems incredulous when viewed through the prism of history, but doubtless they had their reasons. Attempts to find out how or why these changes took place have been unsuccessful and a request to the City for written minutes of this time have met with a "there are none" response. For this reason, the 1970s dark period of the Garden's history has had to be described more speculatively than a true researcher would have liked.

So, Helen Connery's worry that there could be future puzzlement over the development of Mead Botanical Garden became a reason for this book. It is an attempt to correct the misconceptions and misinformation, particularly about the early years of the Garden before the City took over, provide factual knowledge as to dates and events, and document as true and accurate an account of the development of Mead Botanical Garden as possible, including the vital role played by Jack and Helen Connery.

Paul Butler
Winter Park
May 2019

PROLOGUE

The Opening of the Garden

On Saturday, January 14, 1940, the Mead Botanical Garden formally opened its doors to the general public. It had been a three-year struggle, full of laborious work and financial juggling, to reach this point and create such a beautiful legacy garden out of the no-man's land mixture of jungle and wetlands located in the south-west corner of the City of Winter Park, jutting out at its southern point into the City of Orlando. The two chief architects of this botanical vision were Jack (John) Connery, youthful admirer and disciple of the famous botanist the Garden was named after, Theodore Mead, and Edwin Grover, Professor of Books at nearby Rollins College, who had worked for years to acquire property and obtain backing for the Garden. Now, with happiness and great satisfaction, they could both look forward to this day of pageantry and celebration.

The day dawned warm and bright with a brisk breeze cooling the nearly three thousand people who came to attend the ceremonies and wander through the spacious 55-acre garden. As the 2 p.m. commencement time approached, parking close to the Nottingham Street entrance began to be jammed with cars and people spilling out into the surrounding streets. Entertaining the crowds before the official opening were the Winter Park High School Band, under the direction of Gene Sturchin, appearing for the first time in a public performance in their stylish orange, black, and white uniforms. Adding a sense of the Old South to the occasion were members of the Orlando Junior Welfare Association, beautifully gowned in Southern belle costumes and wearing exquisite camellias from the Garden in their hair, who would be acting as hostesses to the dignitaries.

Two ceremonies were planned that afternoon, the first at the Orlando entrance, where a temporary installation of radio transmitters had been set up to beam an account of the ceremonies via WDBO radio and announcer Bill McBride. The formal program was due to start at 2:30 until 3 p.m., with brief talks by Rollins College President Hamilton Holt, Mead Botanical Garden President Edwin O. Grover, Senator Charles O. Andrews Sr., Orlando Mayor S. Y. Way, and Winter Park Mayor J. F. Moody, culminating in the main ribbon-cutting formality at the Orlando entrance. Following this, the guests were to walk through the Garden to take part in a briefer ribbon-cutting event at the Winter Park entrance on S. Pennsylvania Avenue.

As the event became live on air, Grover took the microphone to act as master of ceremonies. He first introduced Hamilton Holt, honorary vice-president of the Garden, who paid tribute to Jack and Helen Connery for their tireless efforts, and stated: "When Winter Park and Orlando get behind anything, it is sure to succeed." He continued, "A garden is never completed, and I'm proud that these two cities are behind something which will never be finished but will go on and on." People whose gifts and help made the opening possible were mentioned, among them Senator Walter W. Rose who gave roughly twenty acres of land; James Treat, former Winter Park Mayor, who donated around seven acres; R. F. Leedy, who also gave some property; and Mrs. Mary Bartels, who bequeathed approximately 20 acres for the arboretum, planned for the high pine lands area of the Garden. He also praised Dick Pope who aided with the publicity but who was unable to be present because of illness.

The focus on the opening ceremony was brevity, so none of the speeches was long. Jack Connery took first prize for conciseness when, upon being introduced by Grover to speak over the air, he puffed out his chest and said, "This is a great occasion for all of us," and immediately sat down again. At which point Grover commented, "You see, Jack is a doer, not a speaker."

Senator Charles O. Andrews Sr., speaking briefly, said he had just "gotten home from the frozen North," to find here "one more lovely beauty spot in Florida." "I will continue to work in Washington to give you what aid I can," he promised. Like the others on the program, Mayor Way of Orlando was present when ground was broken at the Garden two years ago and was called upon to speak briefly again. He stated "From this moment on, the garden is a place of refuge for communication with nature," and concluded, "What used to be a city dump, is now a bond of beauty between Florida's two finest cities, Orlando and Winter Park."

As if to facilitate the sinking in of this last comment into the minds of the audience, Bruce Dougherty, the popular tenor of Winter Park, then sang a gay and melodious garden song accompanied at the piano by his wife, after which Mayor Moody of Winter Park was introduced. "Try to visualize what the Mead Garden will be like four or five years from now" he urged. "Only one-tenth of the money spent, and the work done is now visible. It is in the ground and will someday be seen in all its glorious beauty. The Mead Botanical Garden is an outstanding attraction of Central Florida from now on."

With the speeches completed, the prominent guests were paired off with the costumed hostesses, and the group strolled to the gates that were closed with a bow of purple ribbon. Mrs. Claude Pepper and Mrs. Charles O. Andrews Sr., both wearing impressive orchid corsages, snipped the fabric with scissors entwined with orchids and green streamers. With the gate opened, and the band still playing, the rest of the guests entered the Garden and mingled with the dignitaries. Throughout the site, visitors could look forward to viewing one of the rarest and most valuable collections of flowers and shrubs from all parts of the world. But the jewel in the crown for many was the spectacular display of over 1,000 orchids from Mead's original collection, contained in the large newly erected greenhouse adjacent to the Orlando gate, with over a hundred of the most beautiful varieties in full bloom.

Led by Hamilton Holt, and accompanied by commentaries from Edwin Grover, they wandered along the myriad peaceful woodland trails that crisscrossed the Garden and in places followed the banks of the stream running through the property, each path covered with sand and cypress mulch to facilitate walking. They marveled at the intriguing vistas of mirror pools and remarked on the beauty of the low waterfalls of the cascading creek. They heard and saw Grover point out a formal piece of land, where 15,000 tulips from Holland had been planted in four squares, around a plot planted with hyacinths, and vowed to come back in a few months when they were in bloom.

Along the brookside trail towards the Winter Park entrance, they gazed at the delicate azalea bushes now just coming into bloom and admired the many types of palms planted along the creek. Grover reminded the group that there were no palms at all when they first started, and Jack Connery had planted them all; nearly a thousand. At the Winter Park greenhouse, erected over the winter to contain Mead's orchids before they were moved to the main greenhouse, they saw many rare ferns, bromeliads, and several thousand coral berry plants which had been grown from seed. Hundreds of cuttings of pink poinsettia, blue hyacinth, bronze bougainvillea, and more than forty varieties of crotons, secured from some of the finest estates in Miami and Coconut Grove, were on display in an adjoining slat house.

On the way back, a side excursion took them to the tiny, hidden Lake Lillian where herons and egrets nested and where some of the guests might have been lucky enough to catch a glimpse of the resident alligator on the far shore of the lake. Overall, there was a general appreciation of the tremendous amount of work done to convert this swamp jungle into such a beauty spot, and an anticipation of a still-richer future to come once all the many plantings were in flower. When it was time to go, as attractive mementoes of the occasion, the guests were given automobile stickers and bumper strips featuring a depiction of a *Cattleya* orchid and the words, "Visit Orlando and see the Orchids."

The future seemed bright to Edwin Grover and Jack Connery. Together they had created the nucleus of what promised to be one of the most beautiful gardens in America's south. The Garden was Jack Connery's brain-child, but he couldn't quite believe what had been accomplished from the day he sought out Edwin Grover in his Rollins College rooms and shared his concern about what to do with all the orchids and other plants that Theodore Mead had bequeathed him. Now it was a reality, but right at the beginning, just after Theodore Mead's death, there was uncertainty as to whether he could find an appropriate location; or even whether he had legal ownership of any of the plants needed for a memorial garden.

CHAPTER 1

How the Garden Came About

The two founders of the Garden, Jack Connery and Edwin Grover, had both known Theodore Mead since the 1920s and were familiar with his estate around Lake Charm in Oviedo. Jack Connery was an Orlando scout when he first encountered and became friends with the Scoutmaster of the Oviedo troop at the Silver Lake scout camp in the summer of 1922. He was attracted to the charismatic scout leader, who was fun to be around and a knowledgeable father-figure. Mead was popular with all the boys and much in demand at camp meetings with his energetic story-telling, his encyclopedic knowledge on many subjects, and, with his strong melodic voice, his leading of the singing of campfire songs. He was fond of augmenting the traditional Boy Scout songs with those he remembered from his fraternity days at Cornell. His party piece was a rendition of *Green 'n' Yeller*, based on *Taranty my son*, a New England ballad about a boy poisoned by his grandmother by being fed snakes rather than eels for supper, which invariably had the boys in stitches with its chorus of fake vomiting.

Throughout the 1920s, the two of them continued to meet occasionally at scout camp meetings and developed a shared interest in ornithology, with Jack having started a birds' eggs collection. In 1930, Connery joined an expedition organized by William Beebe and Otis Barton to conduct exploratory deep-sea descents in their spherical steel vessel, the Bathysphere, off Nonsuch Island in Bermuda. His official role was as cook and assistant photographer and he took many photographs of the expedition and of the strange deep-sea creatures that the ship's dragline brought to the surface. He recorded the events of the first record-breaking dive to 803 feet and the second, on June 11, 1930, to 1,426 feet, nearly three times deeper than any previous diver. Unfortunately, later that summer, he slipped between the *Skink*, Beebe's twenty-six-foot power launch, and the wharf, fracturing a bone in his back and had to abandon the expedition.

Returning to Orlando, Connery renewed his friendship with Mead, and although he had no money, approached Rollins College with a view to studying there. He ended up donating his extensive collection of birds' eggs and nests to the Thomas R. Baker Museum of Natural History at the College in exchange for tuition, and on the understanding that he assisted with ornithology classes and became Student Curator to the museum.

At Rollins College, Connery got to know Edwin Grover and throughout 1932 and 1933 started to take parties of Rollins College students to see Mead's gardens and greenhouse. By this time, Mead had relinquished his official role of Oviedo scoutmaster but still had informal scout meetings on weekends. On one of these occasions, Jack gave a projector presentation on birds, their nests and their eggs.

In 1932, at the age of 24, he borrowed Mead's orchid negatives to make into colored lantern slides and started to document Mead's orchid collection, fearful that the hybridization knowledge would be lost if he didn't. He became a frequent visitor at Lake Charm, helping Mead with horticultural activities such as the repotting of orchids in the greenhouse and the collecting of caladium seeds. Although his museum work kept him busy at Rollins College, he still found time to fall in love with Helen Corey Golloway, a fellow student, who helped out at the museum and shared an interest in natural history and botany. They were married on May 30, 1934, in North Canton, Ohio. Jack and Helen Connery would be the unsung heroes in the early success of Mead Botanical Garden.

1.1: As a young teenager (bottom left), Jack Connery first met Scoutmaster Teddy Mead in 1922 at a Central Florida scout camp (top). Over the years, he became Mead's dedicated apprentice and horticultural assistant, promising his mentor that he would one day create a memorial garden to him. The image on the right shows Mead, circa 1930.

1.2: Jack Connery married Helen Golloway, a fellow student at Rollins College, on May 30, 1934, in North Canton, Ohio. Jack and Helen became essential figures in the early development and running of Mead Botanical Garden.

Mead's declining mobility, now he was in his 80s, was recognized by Connery, who kindly drove him to various events, meetings and on various seaside outings, including one memorable daytrip to Daytona Beach. Theo was invited to dine with the Connery's in Orlando on many occasions, and on February 23, 1933, they gave him a surprise 81st birthday party at their house. Over the years, Jack Connery had formed a robust binding friendship and close master-pupil relationship with the grand old man, and sometime before Mead's death had once said to him in boyish enthusiasm, "Someday I am going to build a memorial garden for you," but Theodore had laughed and modestly brushed the suggestion aside.

Edwin Grover's path to Theodore Mead was a little different. From an early age, he had a love of nature and there was a family interest in botany, focused through Grover's elder brother, Frederick. He became Professor of Botany at Oberlin College in Ohio where the College Herbarium was a national resource for botanists. Both brothers knew and admired Mead's work and Edwin Grover had visited Mead's extensive gardens on a number of occasions. Almost certainly accompanying him on some of these visits would have been his plant-loving brother, Frederick, who on his initial stay with Edwin in Winter Park reportedly said, "I want to meet the famous Mr. Mead."

For Edwin Grover, if the Mead plant collection became available, it could be the nucleus of a valuable botanical resource for Rollins College, similar to his brother's in Ohio. When Jack Connery enrolled at Rollins College and met him, he found they both shared similar aspirations to preserve the collection, and with Mead's passing in 1936 the stage was set for a perfect meeting of minds and dreams – for Connery to create a memorial garden to Mead and for Grover to have access to a botanical garden for the benefit of students and staff of Rollins College. Before any of this could happen, however, there was a need to resolve the legal matter of who owned Mead's plants, and to decide on a suitable location.

Theodore Mead wrote his will in August 1933, two and a half years before he died. His stated intention at that time was to have most of his greenhouse plants go to the Royal Palm State Park in the Everglades, then sponsored by the Federation of Women's Clubs. However, in 1934 Florida suffered a severe freeze which killed many of the plants in the Royal Palm Park, causing him to change his mind over the location, but he forgot to change his will. Compounding the confusion, in various letters over 1935 he promised Jack Connery a representative collection of orchids, caladium and amaryllis, and Clifford Cole of Miami the remainder of the orchids.

The various parties had no desire to split up the collection, and knowing Connery's desire to create a memorial garden, the representatives of the Women's Clubs passed a resolution in June 1936 relinquishing all claims and giving the plants to the executor of Mead's estate, John Augustine Willis. He, in turn, divided them between Connery and Cole and then magnanimously Cole gave his share to Connery. Ownership of the plants was now decided, but the issue of location for a memorial garden was far from certain, and at this stage Connery and Grover were pursuing independent agendas.

Edwin Grover was active initially in trying to obtain Mead's old place in Oviedo for Rollins College, and having been told that there was no money available for such a purchase, wrote to Willis asking whether the heirs of the estate might consider giving the property to Rollins as a memorial. He suggested that it could be known as the "Mead Arboretum and Botanical Garden" and that his brother Frederick Grover might be the Director and supervise the development. This, too, was rejected. Independently, in June 1936, members of the beautification committee of the Orange County and Greater Orlando Chamber of Commerce tried to purchase the tract in Oviedo and designate it as a Florida State Park. They made an appeal for State assistance, which was turned down. Meanwhile, Jack Connery had started exploring property possibilities in Orlando. At his home at 2334 Fairbanks Avenue, Winter Park, he had begun collecting many of Mead's plants, fearful that someone would steal them from the empty Oviedo property. He erected a greenhouse for the orchid collection and had 20,000 amaryllis bulbs laid out drying in the garage. By August 13, 1936, Connery had a promise of a location in Orlando for the "Mead Memorial Garden," at the east end of Lake Eola, down by the old tennis courts.

In early 1937, Edwin Grover was still thinking of ways to try and buy the Oviedo property, when one Sunday afternoon Jack and Helen Connery came to see him for suggestions on how they could take a boat trip in the Caribbean and write up the experience for a magazine. In the ensuing conversation Grover mentioned buying the Oviedo place, but Jack responded that he knew of a better place than that within the limits of Winter Park. He proceeded to describe the location, "You go down to the bridge on Pennsylvania Avenue, and it's that swamp on the right-hand side. There is a brook that runs the whole length of the swamp, and there is a rookery for herons and egrets – but you can't get to it."

Grover was curious to know how Jack knew there was a rookery there if you couldn't get to it. But Jack had the answer, "Any afternoon, go down to Pennsylvania Avenue near the swamp at about five o'clock, and you'll see hundreds of birds swoop down toward a little lake that is there. The birds spend the day over on the coast to feed and return here to roost and nest."

They agreed to go through the swamp and explore the rookery the next morning. Just as the explorers were getting ready, Robert Mitchell, a horticulturist of Orlando and an authority on the Seminole Indians, happened to come by and so the three of them set off. With high boots on and hatchets in hand, they waded and cut their way through the jungle swamp until they came upon the little lake. At the lake, just above where they stood, was a heron's nest with several eggs in it. As Grover recounted, "Jack sat on my shoulders to take a picture of the nest; the first picture taken in the Garden, and it was a beautiful one." They tramped all over the area, found the brook, and decided that this was indeed the place for a botanical garden. Nine Orlando lakes flowed into the little stream, Howell Creek, which ran from Lake Sue at one end to Lake Virginia at the other end and eventually entered the St Johns River. There was a nine-foot drop throughout its length through the property, so there would be an opportunity for several small waterfalls in the Garden.

Jack Connery had done his homework and already knew who owned the swamp – Senator Walter Rose of Orlando. Edwin Grover wasted no time in visiting the Senator at his office that afternoon in the modern Walter W. Rose Building, next door and north of the Angebilt Hotel on Orange Avenue, to ask about acquiring the property.

Walter Rose was a leading realtor and property developer in Central Florida and a State Senator representing Orange County from 1933 to 1949. Featuring the slogan sign, "Rose knows where the money grows" along the main route from Winter Park to Orlando helped him create some of the most beautiful real estates in the area. He developed Orwin Manor, close to the Winter Park line, one of the most prestigious subdivisions of the time and forerunner of other later high-class developments, many carrying his name; Rose Terrace, Rose Hill, Rosemere, Rosearden (a park in Orlando), Beverly Shores, and exclusive Rose Isle, where he built his home. Rose kept horses on his 1,000-acre Fairvilla ranch, helped shape legislation which converted the Everglades into a national park and had a resolute conviction that land would have ever increasing value and contribute to the prosperity of Central Florida. Beautification was important to him too, and the subdivisions he built had attractive plantings of palms and tropical plants. He even started his own nurseries to supply the needs of the developments, and it was said that at one time they contained more than twenty thousand palms.

But Rose was a canny businessman, and if the city, county or state wanted some of his lands for useful purposes, he would generally give it to them, on the basis that it would help him. His reaction to Grover's request for a small swamp from his property portfolio fell into this category. He had built Beverly Shores and bought the swamp just north of it to protect the subdivision so nothing harmful could be built there. From a business perspective, to make the almost twenty acres of swamp lying within the boundaries of Winter Park into a botanical garden was a no-brainer.

But Grover wasn't quite done and asked about including the area between the present boundary line and Nottingham Street, a little over two acres within the City of Orlando, on the basis that it would make sense to have a grand entrance to the Garden available for the people from Beverly Shores. As far as Rose was concerned, congenial surroundings such as access to parks and gardens paid dividends in increasing the value of a property, and he agreed to the request. All parties appeared satisfied, with Grover remarking, "The best afternoon's work I ever did."

If people were donating land for a memorial garden, it needed to be to a legally sound corporation. The charter for the "Theodore L. Mead Botanical Garden," located in Orange County, Florida, was signed into law on May 11, 1937. The elected officers were: Robert Bruce Barbour, President; Raymond W. Greene, 1st Vice-President; Mrs. Charles Sprague-Smith, 2nd Vice-President; John (Jack) H. Connery, Executive Secretary; and Harold Mutispaugh, Treasurer. Connery was the only elected official entitled to receive remuneration, and the purpose of the corporation was stated as being:

> To establish a Botanical Garden and Museum, including greenhouses, an aquarium, an herbarium, for the collection and culture of plants, flowers, shrubs and trees; the advancement of botanical science and knowledge, the prosecution of original researches therein and in kindred subjects, for affording instruction in the same, for the prosecution and exhibition of ornamental and decorative horticulture and gardening, for the entertainment, recreation, and instruction of the people.
>
> To own in its own right real estate for the purpose of developing the things hereinbefore set forth and to secure property and preserve any and all kinds of plant, vegetable, and animal life, thereon. To endow and secure endowment for the perpetual development, maintenance, and care of the things aforesaid. To have the right to dedicate said property owned by the corporation for the perpetual use of the public, to buy, sell, lease, mortgage and exchange any and all kinds of real and personal property as might be necessary in developing and carrying out the aims and objects of the corporation.

The Rose deed was signed on December 1, 1937, formally conveying the two land parcels to the Theodore L. Mead Botanical Garden, a "non-profit corporation organized for benevolent and charitable purposes." As expected, the deed, written by an experienced and leading realtor, was artfully laced with legally binding restrictions and reversion clauses, the principal ones being that it must be developed and permanently maintained as a botanical garden, and the Beverly Shores entrance area could never be built on.

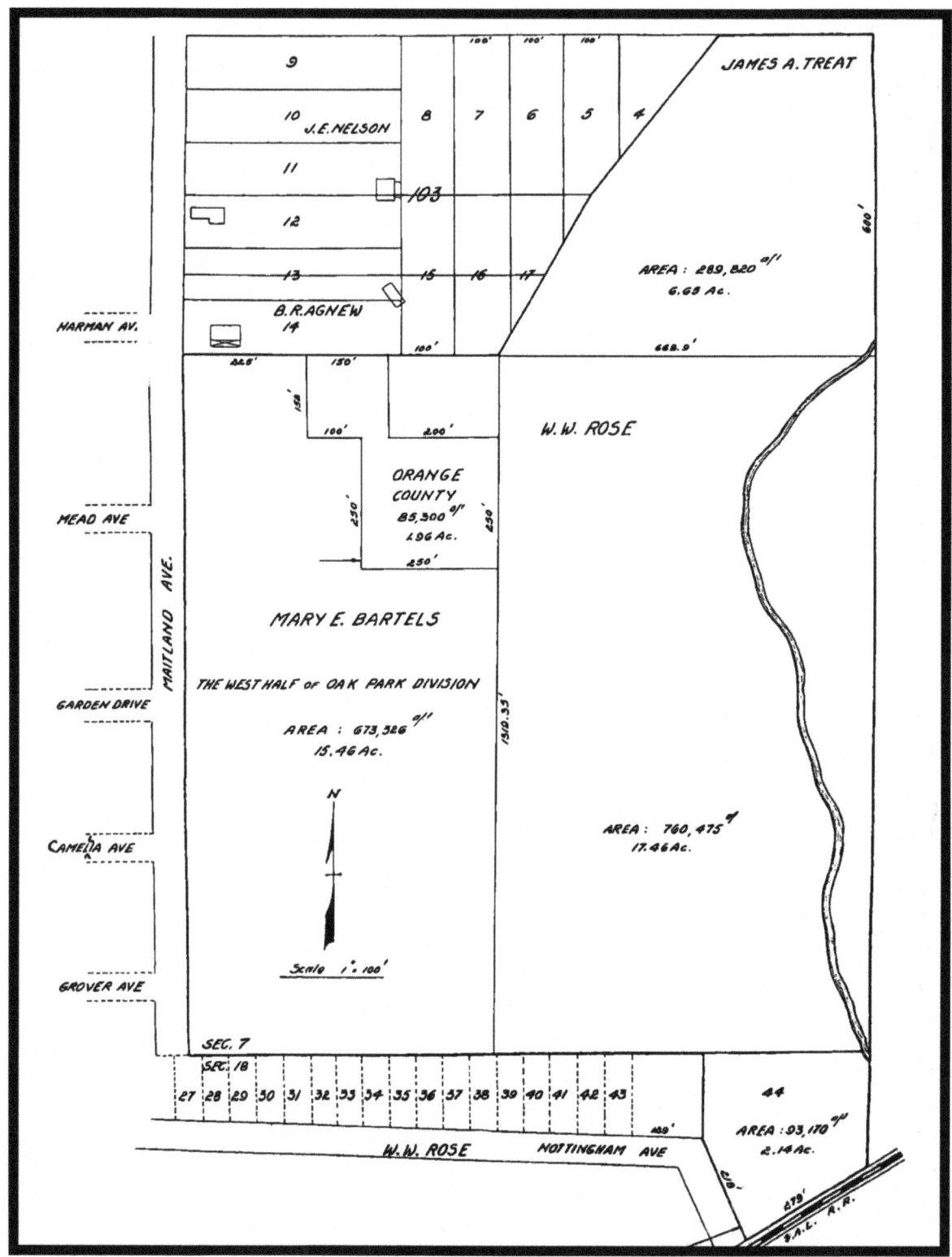

1.3: The original land owners of the property that would form the majority of Mead Botanical Garden were Walter Rose, Mary Bartels and James Treat. The clay pits at the north end of the Bartels tract were originally owned by Orange County before becoming part of the Garden. Not shown is the triangular section known as Cherokee Park, which appears in figure 1.4.

Attention now turned to the swampland area north of the Rose land where Connery had identified the rookery, an area of around seven acres. This belonged to James A. Treat, Mayor of Winter Park from 1933 to 1935. Grover shared his vision of the botanical garden with him and stressed the importance of the small grassy pond on his property that would become an integral part of the Garden and an important bird sanctuary for the City. After a discussion of the area desired, Treat agreed to give the land on condition that it be developed and used as a botanical garden in conjunction with Senator Rose's property, and that the little lake be formally named Lake Lillian, in honor of his granddaughter, Lillian Treat Simkus. The City of Winter Park officially granted his request and on December 25, 1937, issued over the Signature of the Mayor and the City Clerk and the Corporate Seal a document to that effect as recognition of the gift of land by former Mayor Treat.

Treat's part of the bargain, to convert a promise into a legal deed, was slow in coming, and it would be more than two years from Grover's visit before he signed the deed, on June 7, 1939, releasing his land to the City of Winter Park. The deed called for certain conditions to be fulfilled: that the area be fenced to the north; that markers naming the lake and his gift be placed on the lake bank; that there would be a gate in the fence with a path to the lake; and that all parts would be maintained in good repair and in passable condition. In addition, it carried all the usual reversion clauses.

Connery and Grover now examined the area to the west between the Rose/Treat land and Maitland Avenue*, an approximately 20-acre tract of high sandy pineland running from the city line to the north end and enclosing a large claypit area that was owned by Orange County. Although nutrient-poor and not well-suited to growing acid-loving shrubs and flowering annuals and perennials, it was agreed that the entire tract could be a good candidate for an arboretum within the Garden, and so they set about finding the owner.

The property rights of the major part of the land were traced to a Jacksonville

* Now Denning Drive

woman, Mrs. Mary Bartels, whose father, Peter Mack, formerly owned all the area west to Highway 17-92, including what later became the Garden Acres subdivision south of Garden Drive. Connery and Grover made two trips to Jacksonville to see her, not finding her at home the first time. She agreed to donate the land in the west half of the Oak Park subdivision to the City of Winter Park for the use and benefit of that worthy project known as the "Mead Botanical Garden," as recorded in the deed dated August 10, 1938. The relative weakness of the wording of this deed compared to the Rose one when it came to conditions, restrictions and reversion over the use of the land, and the fact that garden development was not concentrated there, would in future lead to claims for alternative land uses outside that of the original deed intentions.

The original claypit tract, 1.96 acres in total in an irregular pattern, consisted of two separate clay-rich zones and lay at the north-west corner of the Bartels plot. It appears that the entire tract was deeded to Orange County by the City of Winter Park to provide clay for various road-building projects, on the condition that the City removed the overburden so that the clay zones would be readily accessible. One of the zones was extensively mined and emptied of clay, leaving squared, perpendicular walls around the four sides of the fifteen or twenty-foot-deep hole; the other remained untouched as the engineers had taken all they needed out of the first zone. Connery and Grover wanted clay and overburden to build up the Garden trails and approached the Orange County Commissioners, who agreed to return the land to the City, which they formally did in a deed dated August 1, 1939.

The Garden was being put together piece by piece and word was getting around, spurring other civic-minded citizens to join in. Robert and Ruth Leedy, owners of a dry goods and clothing store known as "Leedy's" at 311 Park Avenue in Winter Park, agreed to donate a small 0.4-acre parcel of their land adjacent to Howell Creek and the bridge, and on the east side of S. Pennsylvania Avenue. The reversion clause in the deed ran, "If this property should cease to be used for the Theodore L. Mead Botanical Garden, Inc., then this property shall revert to grantors."

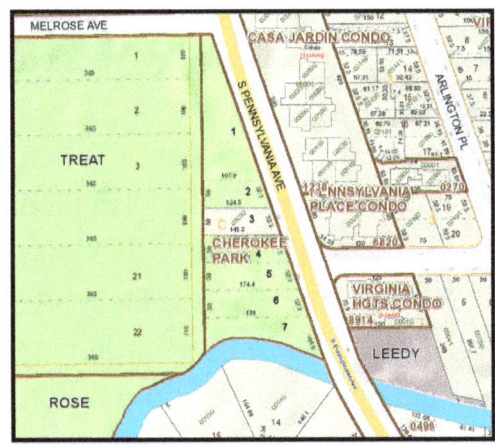

1.4: *Cherokee Park consisted of seven lots which over time were acquired by the City of Winter Park, except lot 3. The Leedy donation of 0.4 acres is also shown.*

To complete the property, Connery and Grover were left with a strange triangular tract of about 1.5 acres, identified as Block C of Cherokee Park and previously known as Block T of the Virginia Heights subdivision. The area ran from the corner of Melrose and S. Pennsylvania Avenue to the Howell Creek bridge, abutting the Treat land, and was made up of seven lots. Connery and Grover realized that they needed a convenient Winter Park access point to the Garden and this could be it, at least until the time when they might acquire the land to the east of Howell Creek. The City of Winter Park already held title to four of the lots on the block, so there seemed no impediment to eventually acquiring the entire triangular tract.

It was time to bring the City of Winter Park on board and try to work out some funding for the venture. Mayor Moody was already an interested party, and he invited Edwin Grover to present his proposals at a regular meeting of the City Commission on Monday evening, April 9, 1937.

At the meeting, Grover described the salient features of the Garden. He went on to say that the plan was only tentative, but if approved and funded, the initial task would be to clean out and deepen Howell Creek to help drain the swamp and widen it in places to create several pools, each ending downstream with small waterfalls to benefit from the elevation change in the creek. Scattered throughout the park would be several azalea gardens with a number of palm gardens, cactus gardens, and an extensive collection of native and imported bamboos. Wooded trails were to connect all these areas and the major structures and provide a means of visiting the rookery and small lake at the north of the property. When completed the Garden would contain more than four miles of beautifully shaded trails, crossing and re-crossing Howell Creek over rustic bridges, with vistas across the mirror pools filled with day-blooming colored tropical lilies. He made the point that the low-lying land needed to complete the area of the Garden, as shown on the map, extended east into adjacent property, and it was hoped that the owner of this tract would also wish to cooperate with the Botanical Garden in enlarging its area, so as to have control of all the low-lying land drained by Howell Creek. Finally, Grover stressed that the operation of the Garden would be closely related to Rollins College, which would be represented in its management. He added that the location of the Garden only a few blocks from the Rollins campus should make it of great service to the departments of botany, biology, and zoology.

1.5: *Aerial map circa 1939 of the proposed Mead Botanical Garden site, with the main entrance on Nottingham Street in Orlando at point A. The expectation was that the land to the east of Howell Creek would eventually become part of the Garden and an imposing Winter Park entrance would be established at point B.*

In the subsequent discussion, not surprisingly, the issue of funding came up. The Grover proposal relied on successfully securing a Works Progress Administration (WPA) grant of around $20,000 which would pay for the labor needed to clear the land, prepare it for planting, and create the trails and pools. The City liked this idea and had no problem in agreeing to put the title of the land temporarily in their name, as one of the WPA conditions. This was done with the understanding that as soon as the Mead Botanical Garden organization was able to take over the property, the city would turn it over to them on a 99-year lease. There was a sticking point, though. In 1937, the City was in poor financial shape, in default on its $100 bonds, which were selling at $39, so it couldn't put up the matching funds required by the government. To the rescue rode Jack Connery, whose thousands of bulbs, rare orchids and hundreds of tropical plants, bequeathed to him by Theodore Mead, had been conservatively valued at $20,000. He let the City put the plants in as collateral in place of the necessary money.

At this point, all of the Commissioners expressed enthusiasm for the project, and it was moved by Commissioner Greene and seconded by Commissioner Baldwin that the City of Winter Park accept title to the property and that Mayor Moody be authorized to appoint a committee to proceed with the application for a WPA grant. This motion was unanimously carried. Greene was named the chairman of the committee on account of his experience in connection with the development of Highland Hammock at Sebring. Mayor Moody was emphatic in stating that immediate action should be taken and the matter pushed forward as rapidly as possible, adding that he believed that when completed the Mead Botanical Garden would attract thousands of people each year to Winter Park.

In early April 1937, Connery and Grover were leaving through the Winter Park entrance, having cleared the first trail, when along the path came a man with two little girls by the hand, who called out, "Who in the devil is in back of this? It's the best thing that ever happened to Central Florida!" It was Martin Andersen, editor-owner of the *Orlando Sentinel*, who then returned to the paper and wrote a three-column piece in praise of the Garden. Throughout his time at the newspaper, he was a fervent supporter of Mead Botanical Garden, using the medium of print to inform Central Floridians of upcoming events and encourage them to visit, and, most importantly, to help raise money for materials and structures in the early stage of the Garden's development.

Over the summer of 1937, once the WPA papers had been turned in, Jack Connery and Edwin Grover asked Robert Mitchell, who had made the first exploratory trip into the Garden with the two of them, to draw a plan of what the Mead Botanical Garden might look like.

Mitchell was a born horticulturist with tropical plants part of his DNA. He was the stepson of Henry Nehrling of Gotha, a life-long friend and horticultural partner of Theodore Mead. He had attended Rollins College and graduated from the famous Missouri Botanical Garden in St. Louis, where he spent several years as assistant superintendent in the landscape division. With his extensive botanical garden experience, he was the ideal choice to turn what was in the mind of Connery and Grover into an artistic visualization. Not all the promises for land donation were secured at that time; in particular, the land to the east of Howell Creek. Consequently, the resultant Mitchell map was more an aspirational vision, having many of the features that Connery had always talked about; a marine aquarium, an aviary spanning the brook, and a spectacular greenhouse to contain not only the bulk of Theodore Mead's orchid collection but many palms, ferns, and other rare tropical plants. Howell Creek would be widened and deepened in parts and three large mirror pools created, each ending downstream with small waterfalls.

Under the proposed plan, the Orlando entrance would be from Beverly Shores, and the Winter Park entrance would be from S. Pennsylvania Avenue, initially via a small trail entrance through the currently vacant lots at the north of the site. There was a universal and robust expectation that future acquisition of the fertile east side of Howell Creek would be possible and would provide vehicular access for Winter Park residents through an impressive entrance off S. Pennsylvania Avenue, framed by the aquarium.

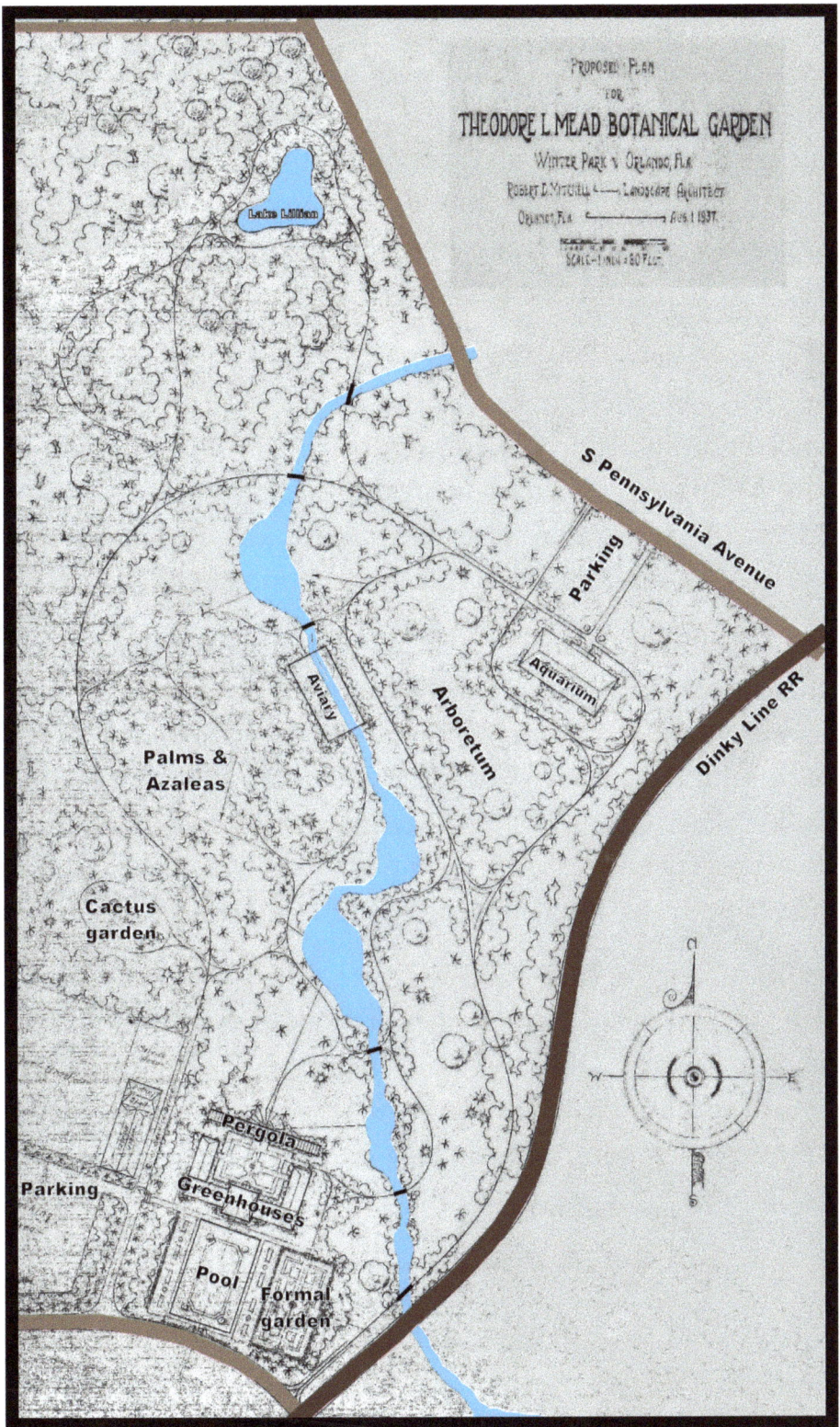

1.6: *The initial master plan for the Mead Botanical Garden drawn by Robert Mitchell, August 1, 1937, showing the imposing Orlando entrance, the aviary spanning the creek, and the aquarium as part of the expected development of the eastern section.*

How the Garden Came About

By late October 1937, with the support of Senator Charles Andrews and Congressman Joe Handricks and WPA officials in Tampa and Jacksonville, the grant papers were in Washington for final approval. In the grant, educational aspects were stressed as the primary benefits to the community, rather than the beautification advantages of a botanical garden. Meanwhile, negotiations got underway in Winter Park to try to acquire the adjacent 35 acres, east of the creek and north of the Dinky Line.

On December 19, 1937, the Project Control Division of the Washington WPA offices announced that funds of $20,170 had been allocated to begin the work, and indicated a willingness to supplement the fund as work progressed. Congratulations in letter form came from several sources, one from Hamilton Holt of Rollins College to Mayor Moody was typical in sentiment:

> There is no doubt in my mind that you and your associates have begun in the Mead Botanical Garden something of tremendous value to the State of Florida – something that will be far-reaching in its effect upon our education and recreational life. Its value to science is beyond estimate. I believe the Garden will in time come to be recognized as an outstanding attraction in Florida to tourists, and I feel sure it will provide a recreational center of charm and genuine value to all our people.

Edwin Grover was feeling elated and set a date in early January for a groundbreaking ceremony, and asked Jack Connery to organize a WPA labor force that could start work in January 1938. Between them they had decided on an aspirational slogan to sum up what they were trying to achieve, and arrived at "To be Florida's finest garden spot," which they had printed on signs close to the two entrances.

Groundbreaking took place on Saturday, January 8, 1938, by what was to become the Orlando entrance to the Garden. The ceremony, presided over by Edwin Grover, featured Rollins President Hamilton Holt as principal speaker,

Robert Bruce Barbour and Jack Connery of the Mead Botanical Garden Association, and invited dignitaries: Senator Charles Andrews, Senator Walter Rose, Orlando Mayor S. Y. Way, Winter Park Mayor J. F. Moody, Martin Andersen publisher of the Orlando Sentinel-Star, R. F. Leedy, Carl Jackson, chairman of the Orange County Board of Commissioners, and various WPA officials. Grover read telegrams, Holt lauded the work of Grover and Connery in bringing about the project, Senator Andrews stated that today marked the first steps in creating a wreath of beauty in Central Florida, then, taking the spade of gold and silver from Grover, he turned the first shovelful of dirt.

1.7: Groundbreaking day, January 8, 1938. From top left, clockwise: Senator Andrews turns the first spadeful of earth, watched by left to right, Way, Barbour, Jackson, Grover, Leedy, Apgar, Connery, Thompson, and Moody; Setting up with radio station WDBO; Main entrance sign; Winter Park Mayor Moody asks presenter Edwin Grover a question.

Grover then said: "We are here today to honor Theodore L. Mead and to turn the first spade of earth of what is to be a beautiful memorial to the man and his work. Probably every garden of any size in the State of Florida contains at least one plant which Dr. Mead has developed. He was the first to obtain a pure white amaryllis, he changed the colors in crinums, caladiums, daylilies, and many other well-known flowers, and he developed entirely new varieties of orchids."

The ceremony continued with Grover describing in detail the vision for the Garden and the work which would be starting soon to establish several mirror pools and waterfalls along Howell Creek. He added that the land would need to be graded in places, azaleas and palms planted, and nearly five miles of footpaths built, together with suitable entrances at the Orlando and Winter Park boundaries to the property. The work would also provide adequate protection for the many rare birds, including a rookery of herons and snowy egrets which was now part of the garden site.

Laborers hired under the WPA grant arrangement and supervised by Jack Connery and a WPA official started work on January 17, 1938. On average around 50 men were employed each day, cutting trails, clearing underbrush, filling low spots, building pools, widening and deepening the creek, planting the first palms and plants, and constructing dams. By mid-February, the first trail had been built up with clay brought in with wheelbarrows and then surfaced with a thick layer of cypress mulch, and the Beverly Shores entrance had been regraded to the level of the road, necessitating a landfill of about two feet. The sand and clay materials for the trails and landfill were dug out of the large claypit using a loaned Orange County dragline.

By mid-April, two acres of swampland had been drained and cleared for a nursery. One acre had been freed of roots and built up into long beds, with irrigation ditches between, ready to receive plantings of azaleas, ferns, caladiums, and palms. The following week, work started on putting up the two miles of fence around the Garden. The fence posts of cabbage palm trunks were cut and brought in from a swamp near Oviedo. The next task would be to deepen

Howell Creek by around three feet and enlarge areas along it to create lily ponds. Close to the Orlando entrance an oblique pool approximately sixty feet wide and three hundred feet long was planned. Connery wrote to Senator Andrews with a progress report, finishing with, "It is three months today since we started with the WPA and we feel that a great deal has been accomplished, although much more remains to be done."

There was some debate in the community over why the WPA funding level for the Mead Botanical Garden was so low, given that a 20-acre beautification project in Jacksonville recently obtained $77,000 in WPA funds. The stark reason for this was that the WPA was a relief program, designed to feed hungry families of out-of-work people caught up in the Great Depression, and there were simply not as many people in this condition in Winter Park as there were in Jacksonville.

The usual rules associated with a WPA grant application were that the local sponsor provided land, materials, and supplies, with the WPA responsible for wages and for the salaries of supervisors, who were not on relief. However, there were strict restrictions on the amount of time each worker could work, limiting working hours to no more than seventeen days a month, equivalent to around three days' work a week, and consequently slowing down work and spreading eight months of actual work over a period of a year. On the positive side, WPA officials had signified their willingness to supplement this fund if the work progressed satisfactorily.

Materials and supplies could not be purchased with WPA money, so there was a reliance on the generosity of local organizations who donated items of machinery, lumber, tools, and plant material. Putting gas in the truck used to haul in plants and shrubs was frequently down to Connery himself. Throughout the development of the Garden, both Grover and Connery when necessary spent their own money on materials and supplies, amounting in Connery's case to over $3,000.

1.8: *WPA activity in the Garden. From top left, clockwise: Surveying the site; Carrying water for the workers; Steam shovel digging out clay; Deepening Howell Creek; Digging the beds prior to planting; Erecting the palm fencing.*

Connery and Grover were running out of money fast, their own as well as that of the Association, and came up with the idea of printing and issuing revenue certificates, in denominations of $25, $50 and $100, and multiples, bearing interest at 2% per annum. Dated February 1, 1940, and maturing ten years later, they would be payable out of a fund created by reserving 25% of gate receipts. With this printing press collateral and backed by their enthusiasm, they were able to persuade merchants to allow them to buy things they needed, such as gasoline, lumber, and wheelbarrows, in exchange for certificates.

Good progress at the Garden was being made, but it was clear that the WPA grant money would be used up by the Fall, so in July 1938 a supplemental WPA application for a further $42,000 was submitted, approved by the City of Winter Park and Senator Andrews. Even with income from the revenue certificates, there was an urgent need for money to meet incidental expenses – they needed to employ several trained men to prepare and plant the cuttings, azaleas and other plants in the greenhouse; a new garden truck to collect palms and other plants; and fuel for the steam shovel. Optimistically, Connery and Grover expected the Garden to be open in early 1939, with the target date of February 23, the birthday of Theodore Mead. Little did they know that the actual opening would be about a year after that, and they were about to enter their own frustrating winter of delays and discontent.

It all started when Robert Dill, WPA State Administrator, located in Jacksonville, Florida, re-read the various deeds for the property his men were working on, and immediately called an end to all activity. His first problem was with the original Rose deed of December 1, 1937, and the many restrictions it contained, in particular the paragraph stipulating reversion unless "work has proceeded with due speed and has been brought to a state of satisfactory completion." It appeared to make Rose the sole arbitrator in defining completion, since it continued "completion to be deemed to have been accomplished when said property shall have been cleared of unsightly underbrush, weeds, and noxious shrubs, and shall have successfully planted a sufficient number of plants, flowers, shrubs and horticultural exhibits as to render the property attractive and beautiful, and with appropriate facilities for its use by, and inspection of, the public and the same shall be completed in a substantial manner with reasonable speed." Dill wanted the deed changed to eliminate this paragraph, and indicated that work would be stopped until it was. Rose was out of town, but he was eventually tracked down, and agreed, after some argument, to modify the deed. All this took time, however, and it was not until December 7, 1939, that the Rose deed was rewritten.

A WPA crew had just finished creating the Fern Trail leading to Lake Lillian, and out of this came the second and major problem: James Treat still owned the land the men had worked on, and he had not yet deeded it to the City. Again, Dill insisted that work could not continue until this was corrected.

Treat had been central to all the plans for the Garden and specifically to the land he had promised. In December 1937, the City had met his first requirement that the little lake on his property be officially named Lake Lillian. In January 1938, he was included in the guest list at the ground-breaking ceremony but unfortunately was ill on the day. Treat was directly involved in setting the fence lines of his property with the City surveyor and WPA engineer but, despite this close involvement and his repeated statements that he was going "to give the property to the City of Winter Park," he had persistently refused to execute the deed.

Connery and Grover wrote to Mayor Moody in May 1939 pointing these things out and blaming Treat for the six-month delay in returning WPA workers to the site. They added that for the last six months Treat had constantly increased the conditions under which he would deed the property, for example by asking to have fourteen loads of clay hauled onto his property to build a tennis court, which was done at the Garden's expense. Treat also criticized the work of the WPA engineer and superintendent in filling the Beverly Shores entrance and the construction of the nursery, neither in any way connected with his own property.

The letter finished with, "We respectfully suggest that the City Commissioners authorize its City Attorney to take such steps as are necessary to secure Mr. Treat's fulfillment of his agreement, and the delivery of the deed."

Treat was a master of procrastination and intent on wringing every last ounce of property concession in his favor before delivering the deed. Conditions continued to pile up that Connery and Grover had to agree to: that a marker be erected at Lake Lillian stating "This Property the Gift of James A. Treat"; that there would be a gate in the fence along Melrose and a path connecting to other trails in the Garden; and that he and his heirs would have free access to the Garden via this gate at all times and in perpetuity.

When June came around and Treat had still not exercised the deed, Grover must have been close to the limit of his patience. He wrote again to Mayor Moody and diplomatically stated his frustrations, "If we could satisfy the WPA as to the title on the Rose and Treat tracts, I am sure the garden would move forward rapidly and become the asset to the city which it was intended to be." Moody's intervention finally had an effect, and Treat signed the deed on June 7, 1939. Between Rose and Treat, the development of the Garden was delayed by around eight months, and things only started getting back to normal around the end of May 1939.

By this time, the second WPA grant application had been granted and the months ahead were marked by feverish endeavor targeted toward a January 1940 opening date. WPA labor was available, and Connery was ready to start hauling in the many large palms that he wanted for the Garden, but the current vehicle they had, a small Ford half-ton pickup, was not up to the job. Undeterred, Connery and Grover picked up some revenue certificates and went to see Edward Newald, President of Newald Motors, located in Winter Park just around the corner at 666 N. Orange Avenue. They told him about the Garden and how beautiful it was going to be, and how they needed a truck to transport several hundred tall cabbage palms from Oviedo to the Garden.

Newald listened and agreed to help. He had the right kind of vehicle as a used truck; a large twenty-eight-foot Federal semi-trailer type with removable trailer sides, for $1,200. He made the price $750 and agreed to put on new tires, overhaul it completely, buy the license for it, all in exchange for a revenue certificate. He brushed away Connery and Grover's thanks, saying "Don't thank me. I'm not giving you a thing. You're doing something for me. My business is here. My home is here. I get sick and tired of sending people out of Orlando to see Silver Springs, to the Bok Tower, to Marineland, to the Cypress Gardens in Winter Haven – sending them everywhere in Florida to see something except right here in Orlando. You people have a vision and it is a constructive, selfless one for the people of Orlando."

Within days the truck was in use, carrying from Oviedo over a couple of months or so a total of 140 cabbage palms. It was a day's work to go to Oviedo, dig and bring in five palms from twenty to forty feet in height, some weighing a ton apiece, and then plant them. Connery constructed a derrick and palm hoist arrangement with material donated by Transit-Mix, and Woods and Simmons Woodyard. He planted some palms at the Winter Park entrance, others in the twenty-acre arboretum, and the rest down by the waterfall below the lily pool at the Orlando entrance. A further trip with the truck was planned down to Deep Lake in the heart of the Everglades to pick up a load of large Cajeput trees.

Over the summer Edwin Grover had to go to New York City for two months helping to promote Rollins College at the Florida Pavilion of the 1939 World's Fair. Jack and Helen Connery arranged to drive him to Jacksonville where he could catch the train. Before that, the plan was for the three of them to visit local growers and nurseries in the Jacksonville area, to try to secure plants for the Garden. The first port-of-call on Saturday afternoon was the Glen Saint Mary Nursery where they talked with the owner, Lars Taber. He told them that he had a surplus of ten-feet tall evergreen specimen plants, which were too expensive for the average purchaser, and he would be glad to give fifty of them to the Garden if they could be picked up.

1.9: Jack Connery and his crew planting palms trees, circa 1939.

On Sunday morning, at Fernandina Beach, Mr. G. G. Gerbing, who was a large grower of azaleas and camellias, offered to present to the Garden a collection of 2,000 azaleas, containing around 1,400 named varieties. While Connery had thousands of little azalea plants growing in the garden nursery, he welcomed this contribution of large-sized foundation plants that would bloom the following January when the Garden was due to open.

Also on Sunday, they visited Bruno Alberts of Mandarin, near Jacksonville, a commercial orchid grower who specialized in *Cattleya* orchids for the corsage business and who had promised to give an assortment of his choice orchids to add to Mead's collection at the Garden. They saw blooms five to six inches across, and a few at seven to eight inches, all in beautiful colors, and arranged with Alberts that in about a year they would be ready to receive them. It was the end of a perfect weekend. Helen Connery, in describing it to a friend, wrote, "Dr. Grover left on the nine o'clock train more encouraged about the future of the Garden than he had been for some time."

Back in Winter Park the Connerys had some good news concerning their quest for a large greenhouse to be located at the Orlando entrance. The owner of one of the most beautiful gardens in Orlando's Edgewater Heights, John Anthony Porter, had unfortunately recently passed away. A keen horticulturist with an impressive garden display of camellias and azaleas, he had just installed a $1,200 greenhouse but died before he had time to plant it. Mrs. Porter was offering it to the Garden, like new, for $350, but their immediate problem was raising the money.

The financial reality was that they could not afford the Porter greenhouse, neither could they start picking up the plants they had been promised until they had money for gasoline; and they still needed to pay a team of qualified people to transplant the thousands of plants they had been given. The money cupboard was bare and they were living on hope, but over a thousand miles away at his summer home in Annisquam, Massachusetts, was a knight in shining armor who had just asked the Mead Botanical Garden Association for a revenue certificate for $1,000.

The knight in shining armor proved to be the philanthropist Dr. Eugene Shippen. He and his wife, Elizabeth, had arrived in Winter Park in the fall of 1930 and started looking for a home somewhere close to Lake Maitland. He was born in Worcester, Massachusetts, a Harvard graduate and Unitarian minister who served the Second Church in Boston until his retirement in 1930. As another wealthy, cultured New Englander, he was rapidly assimilated into the community and soon met and became a friend of Irving Bacheller, one of Winter Park's most famous residents, and through him met Dr. Hamilton Holt, President of Rollins College, Annie Russell, and other influential Winter Park people.

As luck might have it, there was an empty building plot adjacent to Bacheller's estate, Gate o' the Isles, at 1200 Park Avenue North that Shippen acquired and then asked James Gamble Rogers II to design them a home, to be named "Casa Felice." A précis of the groundbreaking event reported in a newspaper of 1931 gives a flavor of how the rich and famous handled a social event like this at the time:

> Not every prospective home builder can assemble a nationally known educator, a famous actress and noted writers to break ground for the building. That is what happened yesterday afternoon at the home of Mr. and Mrs. Irving Bacheller in Winter Park when friends of Dr. and Mrs. Eugene Rodman Shippen assembled to turn the first sod for the new Shippen home.
>
> The group proceeded to the new Shippen lot by twos. The line was headed by Miss Dean of Orlando who played a charming dance tune on her violin. Arriving at the house site, Dr. Shippen read some humorous verses for the occasion and introduced the speakers. Mrs. Shippen turned the first sod as the representative of the home. Mr. Powers spoke in Italian. Mr. Gamble Rogers II, who designed the house, also spoke. Mr. Irving Bacheller represented the neighborhood and read a beautiful original poem, "Song Shores." Dr. Hamilton Holt of Rollins College made a witty, impromptu talk. Dr. Shippen then called on the arts, and Miss Annie Russell, the well-known actress, spoke briefly but humorously.

The Shippens quickly became part of the elite cultural scene of Winter Park and his new home at 1290 Park Avenue North a favorite venue for social gatherings, with Holt and Bacheller regular guests. Through Holt he met Grover, who also entertained members of the Fortnightly Club at his home on Osceola Avenue, and in April 1932 he asked Shippen to be their speaker for the evening.

In 1939, Eugene Shippen conversed with Robert Bruce Barbour, then President of the Mead Botanical Garden Association, at a meeting of the Florida Historical Society. They both lived in Spanish-styled homes designed by Gamble Rogers II, so enjoyed the talk later in the afternoon about the Spanish discovery and influence on Florida given by Holt at his home on Interlachen, that was attended by close to 100 people. Both Mr. and Mrs. Shippen were members of the North End study circle of the Winter Park Garden Club, and in 1939, Elizabeth Shippen was elected chairman of the Beautification Committee. She reported at their annual meeting in April that they had placed 24 tubs of flowers and ornamental shrubs in front of stores and offices on Park Avenue.

Through the Shippen's contacts with Holt, Grover, Barbour, and the Winter Park Garden Club, they were well aware of the development of Mead Botanical Garden and at times its perilous financial state, as too was Irving Bacheller. He was a Trustee of Rollins College, one of the founding members of the University Club and the person most responsible for bringing Hamilton Holt to Rollins as President in 1925. As a result he and Holt had a close relationship and in April 1939, following an invitation from Holt, he hosted a "Forest Tea" in the nearly-ready Mead Botanical Garden, where among the guests were Holt, Grover and the Connerys.

Bacheller was poetic in his praise for the Garden, and the need for man to connect with Nature. When asked by a newspaper reporter in early May 1939 whether he had any parting words before he headed off North, he cautioned that one thing for growing cities to remember, with their spreading pavements, was the need to keep in touch with Mother Earth. Throughout his life, he said, the beautiful, wild immensities of Mother Earth had been a rich source of inspiration for him. He continued:

> I have discovered a wild and most interesting area of Mother Earth between Orlando and Winter Park. It is that wooded section stretching toward a mile in length between Lake Sue and Lake Virginia with a beautiful brook flowing in forest shadows from one lake to the other. The remarkable collection of flowering plants, the result of years of intelligent effort by the late Mr. Mead, are now growing there.

1.10: Three influential figures in the development of Mead Botanical Garden. Left to right: Irving Bacheller, who at an early stage offered fine poetic words of support; Eugene Shippen, who gave money and with his wife, active practical support; and Martin Andersen of the Orlando Sentinel newspapers, who through his medium of print, regularly publicized the Garden and helped raise money for its opening.

With these fine and generous words, Irving Bacheller left for New York. Meanwhile, in Winter Park, the Connerys could not believe the $1,000 check they had just received from Eugene Shippen. This money would kick-start the plant acquisition and planting program, and suddenly anything seemed possible. Grover returned from his two months in New York at the end of July and immediately wrote to Shippen, expressing sincere gratitude for the gift, the money coming in "just at the moment when help was most needed." He continued by saying that rapid progress was being made and that they were more certain every week of a

January opening. He finished with, "Jack and I were both deeply touched by this expression of practical helpfulness." Connery was equally grateful, writing, "It was your generosity that again started the pendulum on the upward swing when we were almost afraid it might stop. You not only gave the Garden the push it needed but gave us the encouragement and moral support we needed."

With money in the kitty and the WPA men back at work, there was still much to be done in the Garden. The ground had to be cleared, and beds had to be prepared before the many plants they had been promised could be brought in. A crew of men cleared space for the camellia garden and had to dig it over several times because of the matted mass of roots. Areas for the azalea and gardenia gardens were cleared and grubbed. They had decided to change the region initially designated on the Mitchell plan as an outdoor amphitheater into a terraced garden of annuals, sloping gradually down to one of the mirror pools; so this two-acre site had to be prepared as well. Multiple trips were made to the Gerbing Gardens in Fernandina, bringing back around 2,000 azaleas, and to the Everglades Nursery at Fort Myers, hauling back Royal, Cocoanut, Fishtail, Cane, Roebelenii, Rupicola and Washingtonia palms. Also, the Sanford Garden Club had expressed an interest in sponsoring a gardenia garden and had presented the Garden with a $25 gift to start the ball rolling. With this, Connery negotiated the donation of about 100 Veitchii-type gardenias to which the nursery added around 50 of the Belmont variety, on the understanding that cuttings from this patented variety would not be given away. All the azaleas, camellias and gardenias were cut back and planted in the nursery awaiting completion of their beds. Connery had a gang of nine men working for him doing the cutting and transplanting, and about forty WPA men were working on other projects.

Most of the WPA workforce was involved in bed preparation and the hauling and planting of palms. Building an eight-foot clay driveway winding through the palms and pines of the arboretum tract was another task, as was bringing in eight tons of lime rocks from a site fifty miles away. Connery used these rocks as decorative edging for the flower beds and placed some of them at and around the upper waterfall, which Grover described when it was finished with foliage planting as, "with the three leaning palms and the beautiful little island, makes a picture that is already lovely."

1.11: Creek and trail scenes in the period before the excavation of the two lily ponds in 1942. Two of the three large mirror pools ending in small waterfalls are shown.

By early September, the half-ton Ford truck that they had been using for the last year had given out entirely and was not worth repairing. The windfalls of the summer had been used up, and they had no money for a new one, so Connery and Grover went to visit the Heintzelman used-Ford dealership on W. Livingston between Orange and State and asked whether they would accept a revenue certificate, due in ten years. They came away with an excellent used truck, only a few years old, with curtain sides to cover the back and protect the plants from the wind during transit. On top of that, Bob Heintzelman added 1,000 gallons of gasoline for the truck on the same revenue certificate. In need of hardware for work in the Garden, they called in at Bumby's Hardware on West Church Street, who agreed to give them some wheelbarrows and the hand tools they needed against a revenue certificate of $130.

They decided to concentrate on the two entrances for the next two or three weeks to give it a finished appearance. At the Winter Park entrance they planted azaleas around the palms, laid some grass, and erected a fence. Near the trail entrance they built the upper structure and interior fittings of a small orchid house on the stone foundations that Connery had laid several months ago. At the Orlando entrance, Connery and Grover planned to landscape the lawn outside the fence with shrubs and trees and create a sweet pea display either side of the main entrance, extending in a graceful curve for a distance of about 250 feet. Connery wrote to Eugene Shippen asking him whether he had any ideas for the main entrance design and Shippen sent a sketch that was positively received. They particularly liked the two towering palms acting like sentinels either side of the entrance. For the Winter Park entrance Eugene Shippen had approached Gamble Rogers II, whose proposed design used a pergola of cabbage palms to form the opening.

1.12: *The initial Garden entrances. Top: the Orlando gate just prior to opening, without the gatehouse nor greenhouse, showing the two sentinel palms; Bottom: the completed Winter Park gate, 1940.*

1.13: Edwin Grover poses with Jack Connery on May 29, 1938 on the occasion of the christening of Connery's second son, William Edwin, held by godfather Grover.

Jack Connery, together with his horticultural helpers and the WPA labor force, had completed a prodigious amount of work over the four months from July to October 1939, which he recorded in a memo to the Mead Garden Association. At the bottom of the list, he added, "Prepared ground, laid irrigation tiles and planted 225 feet of sweet peas." Meanwhile Edwin Grover was equally hard at work, at weekends and in his vacation time, promoting the Garden in any way that he could. He gave talks, addressed Directors, broadcast over local radio, and took key business people through the Garden, usually having several appointments a day. He aimed to get the various civic organizations in Orlando to back the Garden and help sell sufficient revenue certificates to bridge the gap between then and opening day.

Connery and Grover worked well together, sharing a common purpose and complementing each other's skills. They occasionally got together socially and Helen Connery was, for a time, Grover's personal secretary. When the Connery's had their second son, William Edwin, christened in May 1938, Grover readily accepted the role of godfather. In Grover's honor, the child received the name Edwin, and this became his preferred Christian name.

In late September 1939, Martin Andersen received a tour of the Garden and was amazed at the progress achieved. Connery and Grover poured their hearts out to him over the financial state they were in, with over $10,000 needed to meet the January opening date. This sum was made up of specific items; for example, they had the Porter greenhouse promised to them but did not have the cash to buy it or move it, and without this primary structure for Mead's orchids at the Orlando gate, the opening would be a distinct flop. Andersen committed to helping them get the money and wasted no time in moving into action. True to his word, the next day he wrote an editorial describing the Garden as "an invaluable asset" and praising the efforts and personal sacrifices of Connery and Grover, "who have given a year of their time to the venture, and tear their shirts every Saturday to meet the $100 payroll." The following day he sent out Charles O'Rork of Dixie Studio, with three assistants, a newspaper reporter and fifteen high school girls, to take pictures in the Garden that could be featured in subsequent editions of the paper. In addition, he said he was going to devote the first ten pages of the Mail-Away Edition in November (36,000 copies) to the Mead Botanical Garden and beginning on January 1 he promised to run a daily streamer across the top of the paper urging tourists to be sure to see Mead Garden.

There was more to come. On the morning of Monday, October 2, 1939, subscribers of the *Orlando Sentinel* came down to the breakfast table, picked up the newspaper, and read a blistering editorial by Martin Andersen concerning the dire financial state of the Garden. The page one headline, displacing news on Hitler and the just-started war in Europe to a secondary position on the page, was entitled, "A Challenge to this Community," and started with his impressions of current life in the Garden:

> Ten negro laborers trudge along pushing Bumby-contributed wheelbarrows, where there should be five times as many at work. Jack Connery, driving an Ed Newald-contributed truck from one o'clock in the morning till dark, bringing in palms from Everglades, azaleas from Fernandina, other plants from all over Florida, did not get a dime for his efforts last week. He is the garden director. The director business must sound pretty high-falutin' to Jack who has been in overalls and behind the wheel of a truck or on the business end of a shovel now for a year. Meanwhile, the erstwhile stiff and staid college professor, Dr. Edwin Osgood Grover, who never before in his life had encountered a payroll has been borrowing, begging, contributing and doing everything but committing theft to keep gasoline in Connery's truck and those negro boys at work.

The editorial went on to berate organizations who were reluctant to get involved and give a hand, including the newspapers which had not aroused the community to the growing financial plight of the Garden, and asked: "Where is our civic leadership?" In his view, "We've all got to put it on the line. The quicker we do it, the sooner and better the job of completing Mead Garden and insuring Orlando's place on the winter resort map."

Over the next few days, the editorials and coverage continued, urging the setting up of a money-raising campaign with someone to run it. The Directors of the Garden agreed to employ a professional person to organize the workers in the various civic clubs in Orlando and Winter Park who would do the legwork in the financial campaign. The Winter Park Chamber of Commerce, the Winter Park Lions Club, and the Orlando Greater Chamber of Commerce endorsed the Garden and offered to assist in the campaign. When Grover addressed various business and local service organizations over the need for financial support, he applauded local companies, such as Newald Motors, Heintzelman's, Bumby's, and Smythe Lumber, who had either readily taken revenue certificates or else donated material supplies. He reserved the highest praise however for the Orlando newspapers, stating in a lunch-time presentation to the Winter Park Kiwanis, "If it had not been for the front-page editorial which the Orlando papers gave us, I don't think we would have gotten far with our project. This past Sunday more than 300 visitors strolled through the gardens admiring the wild beauty and bird life. It is hard to believe that Dick Pope, with his fifteen acres of Cypress Gardens, can attract more than a hundred thousand people every season. Think what we can do with 55 acres."

1.14: *Crowds at the Orlando gate on pre-view day, Sunday, October 29, 1939.*

The newspaper publicity and other promotional activities had generated considerable interest in many Central Floridians, and they were keen to see what the

Garden looked like so it was decided to hold a preview day on Sunday, October 29. The gardens opened at noon, free of charge, with conducted tours under the guidance of Connery and his assistants leaving the Orlando entrance on the hour. The visitors were led through the maze of flower-fringed paths and along the banks of the brook, with a side-trip to Lake Lillian where the trees by the lake were favored by herons and egrets, especially during the nesting season. Then back to the entrance where, for entertainment, the Orlando High School Band of 70 boys under the direction of E. J. Heney as conductor played to the crowds from 1:30 to 5 p.m. In subsequent newspaper reports, the number of visitors was variously reported as between one thousand and five thousand, but whatever the number, it had Connery exclaiming that "The number of visitors gives a small indication of the thousands we can expect after the Garden is officially opened."

On Thursday, November 2, 1939, a city-wide campaign was launched to raise the more than $11,000 necessary to complete the garden development, and on that day the *Orlando Evening Star* carried on page five a five-column wide box, with the title "The Mead Garden Campaign Opens Today – Here is Why this Money is Needed."

Raising $11,513 in 1939 – equivalent to perhaps $200,000 in today's money – in an environment where Americans were slowly emerging from the Great Depression and were now facing the market uncertainties associated with the war in Europe, was a daunting task. To be successful, Andersen decided on two things – that the appeal across the readership area of the Sentinel should stress helping the community, and that it should focus on the Garden's unique selling point, the rare and beautiful orchid.

In Andersen's view, the region could not be sold to visitors by oak trees and silvery lakes set in the middle of green grass lawns because those could be found all over Central Florida. Mead Botanical Garden was the answer to the question; "Orlando – why go there? What is there to do or see?" He suggested the marketing slogan, "Visit Orlando and See the Orchids," asking his readers, "Will Orlando follow the orchid to fame and fortune? The world's most expensive flower, the epitome of Nature's most delicate coloring may eventually represent our town to the traveler."

THE MEAD GARDEN CAMPAIGN OPENS TODAY

HERE IS WHY THIS MONEY IS NEEDED

Budget For Opening Mead Garden

1. Build two gatehouses, one at Orlando entrance and one at Winter Park entrance .... $ 625.00
2. Purchase and install John Porter Greenhouse 1,000.00
3. Bring orchids from Jacksonville 50.00
4. Bring plants from Glen Saint Mary's Nursery, from Royal Palm Nursery and from Southern States Nursery. Plant them and make a formal garden back of the greenhouse 402.00
5. Resurface three miles of trails with clay and sawdust 250.00
6. Complete Camellia Garden. This includes bringing 200 Camellias of blooming size from Gerbing Nursery, Fernandina .. 146.00
7. Complete Gardenia Garden 126.00
8. Prepare ground, build flats and raise 10,000 tulips 528.00
9. Prepare ground, raise, plant large display of annuals as well as certain bulbs including calla lilies. This includes making flats, buying fertilizer, etc., as well as the salary of an experienced seed man 503.00
10. Get orchids from St. Louis 300.00
11. Bring in and plant 250 cabbage palms and plant azaleas at their bases 756.00
12. Enlarge pool and do other necessary dredging 126.00
13. Secure colorful birds and place them on hidden lake 126.00
14. Identify and mark trees, shrubs and flowers 225.00
15. Build two public toilets 500.00
16. Secure and outfit office 500.00
17. Garden Director's salary, three months 450.00
18. Landscape Architect's fees, three months, one man full time 400.00
19. Advertising 3,000.00
20. Emergency fund 1,500.00

Total $11,513.00

Sometime ago Dr. Edwin O. Grover President of The Theodore L. Mead Botanical Garden set out an estimated budget of the amount necessary to put the Garden in condition for opening on January first. The amount of his estimates totalled $11,513.

The Executive Committee of the Garden adopted the budget estimate of Dr. Grover as its budget for the expenditure of the funds realized as a result of the campaign. All expenditures will be made only on specific approval of the Executive Committee, and the budget will be followed as closely as is possible.

It is the intent of the Executive Committee that all of the funds raised in this campaign be placed in permanent improvements.

Invest Now—Do Your Part To Help Your Community

1:15: *In October and November 1939, an inspired campaign by the Orlando Sentinel group of newspapers, spear-headed by Martin Andersen, brought in enough money from Central Florida residents and businesses to meet the budget for the Garden to open.*

Not everyone was happy with this Orlando positioning. There were some Winter Parkers who took exception to the Sentinel's rhetoric and viewed the publicity negatively on the grounds that only about 6% of the Garden's total acreage actually lay in the City of Orlando, and once the fence and entrance gate at Beverly Shores were in position this fell to almost zero. Connery and Grover on the other hand were positive about the campaign, believing that without the money it would raise, the Garden would be nothing, and that the end justified the means.

The money-raising campaign got off to an excellent start with a large team of volunteers phoning and cold-calling businesses. Donations from the general public could be deposited at the offices of The Sentinel and in the lobby of the San Juan Hotel at Orange and Central, and by the end of the first day, a total of $1,321 had been donated. On November 5, the *Orlando Sentinel* featured another half-page box, titled "Facts about the Mead Garden." By day eight, the pledged and received donations reached five-figures with a generous contribution of $500 from the Orlando Utilities Commission. The total reported was $10,097 with the group headed by Ed Newald, Chairman of Advance Solicitations, contributing $6,375. Within another few days, the target had been reached, triggering a grateful Grover to write to Andersen, "I want to express to you my personal and official appreciation of the splendid contribution of publicity and personal work which you have given the Campaign. Without your cooperation and that of your newspapers, the Campaign could not have succeeded as it has."

It was full steam ahead again at the Garden, and the Porter greenhouse was purchased and professionally installed adjacent to the Orlando gate. It was set up just in time to receive a collection of 100 large orchids, ready to bloom, and six flasks containing baby orchid seeds in various stages of development, that were donated by the Baldwin Company of Mamaroneck, New York. The small greenhouse by the Winter Park gate was overflowing with over 500 Mead orchids and others that had been donated, and these too were transferred to the new greenhouse. A gift of 36 large orchids in pots from W. A. Manda of New Jersey joined the others, bringing the total orchid collection close to 1,000. The focus was on setting out as many of the orchids that were blooming as possible for display in the main greenhouse on opening day. As a teaser, the Sunday edition of the *Orlando Sentinel Sun* on November 26, 1939, carried a full-page feature on the orchids of Mead Garden complete with six orchid photographs.

The Garden continued to be a hive of activity with nineteen men on the payroll, including Joe Connery (Jack's brother) and Graham Grover, Edwin Grover's son. Connery had taken the young man under his wing, renting him a room at his home, and teaching him the rudiments of horticulture in an echo of the relationship that Jack Connery had himself had with Theodore Mead. There was plenty of work to do – constructing gatehouses at the two entrances, resurfacing the trails, installing public toilets, dredging the pool by the annual garden to enlarge it, and generally getting the Garden neat and tidy for the opening, now projected for January 14, 1940.

Now with a firm opening date, the Directors of the Garden turned their attention to ways to not only make the opening a special occasion but ensure it would be sought out as a visitor attraction in the years to come. Sixty miles or so southwest of them was Cypress Gardens, in Winter Haven, a botanical garden that had opened in 1936, which was smaller than Mead Garden but similar in that it showcased acres of azaleas, camellias, gardenias, and other exotic plants, with the added attraction of Southern belles working as hostesses. Annual visitor numbers were rumored to be in the tens of thousands. Connery wrote to the owners who told him that their 1937 figures showed over sixty thousand visitors and in 1938 they had had about ninety-five thousand visitors with five months yet to go, charging an entrance fee of 35¢.

For advice they approached Dick Pope, the master showman and promotor of Cypress Gardens who agreed to act as Publicity Director in an unpaid role, describing the challenge as, "I think the Mead Garden has infinite possibilities and the job sort of challenges the spirit of adventure and arouses the old fight in me. My Cypress Gardens are on the upgrade. They now surpass any of the stupendous lies I may have told about them in the past. Their beauty beggars description and even my flow of adjectives." He went on, "I have promised your board of directors with a plan for action, a sort of battle program for the first two months of our drive. They have accepted my ideas to make the Mead Garden name famous."

Not to put too fine a point on it, what was really working for Cypress Gardens in attracting many of the visitors was not only the spectacular azaleas but also the pretty costumed girls, and this became an essential element in Pope's strategy for Mead Garden, starting with opening day. For this event, about twenty members of the Orlando Junior Welfare Association were to act as hostesses, dressed in Southern belle costumes and wearing white camellias from the Garden in their hair. It is difficult to imagine the staid and serious Edwin Grover being swept away with this idea as a permanent feature of the Garden, but he went along with it anyway.

1.16: For the opening ceremony on January 14, 1940, members of the Orlando Junior Welfare Association wearing dresses of the antebellum South acted as hostesses for the dignitaries.

The initial order for 10,000 bumper strips arrived a few days before opening. Initially, the preferred slogan was "Visit Orlando and see the Orchids," but Martin Andersen was keen to promote Orlando as the "Orchid City." In a letter to Andersen, Grover expressed the opinion that using the nickname Orchid City for Orlando would be "far more distinctive and more suggestive of beauty and 'class' than the trite and non-distinctive slogan 'The City Beautiful' which, as you have well said, could apply to hundreds of cities throughout the United States." When the Orlando Chamber of Commerce put out one-hundred and fifty billboards all over the state with the slogan "Orchids in Orlando" with a picture of two large *Cattleya* orchids, Andersen seized on this as placing Orlando in a class of distinction, suggesting that "Instead of being the 'Dandelion City', it hereafter becomes known as the 'Orchid City'."

With the turn of the year came heightened anticipation as the Garden opening approached, and a hope that it would proceed smoothly. But what was certain, however, was that without the urging of Martin Andersen through the pages of the *Sentinel*, and the financial contributions made by the residents and businesses of Central Florida, delivered mostly by the efforts of volunteers, the opening of Mead Botanical Garden with the full greenhouse display of Mead's orchids might never have happened.

CHAPTER 2

The Garden Develops Beautifully

Nearly three thousand people jammed the gates for the opening of the Garden in January 1940. They saw the wonderful orchids and a few other flowers in bloom, and left with the anticipation of far more colorful things to come, with the planted dormant annual beds and the emerging azalea and camellia buds. Now, with the passage of a few months, and some spring rain, the azaleas and camellias were in full bloom and the hyacinths and tulips in the annual garden were a great splash of color. The Garden began to live up to its billing of "Florida's Finest Garden Spot."

The large azalea garden near the Orlando entrance, made up of around 2,000 bushes of about seventy-five different named varieties and donated by the Gerbing Nursery in Fernandina, was a colorful carpet ranging through whites, pinks, reds, purples, oranges, and lavenders. The most remarked upon were the dwarf Kurume evergreens: 'Christmas Cheer', a mass of tiny brilliant red flowers, 'Double Apple Blossom' and 'Coral Bells', beautiful and dainty pink, the pure white cultivar 'Snow', and the tangerine-orange and yellow native deciduous flame azaleas (*Rhododendron calendulaceum* and *Rhododendron austrinum*), discovered by the great botanists and explorers John Bartram and his son William Bartram in American forests in 1765.

2.1: *In the early years, some of the plants that John Bartram and his son William had discovered were represented in the Garden. The azalea garden had vast swathes of the flame azalea (top) and by the Winter Park entrance a Franklin tree flowered (right). This tree was donated by the Franklin Association of Florida, and the lower image shows Mrs. Millar Wilson (standing), president of the association, at the planting of the tree in February 1939, performed by Hamilton Holt (left, with spade in hand). Meanwhile, Edwin Grover (sitting, extreme right) examines the cultivation instructions for this rare and unusual plant.*

The Garden had another connection with the Bartrams. On February 9, 1939, a ten-foot Franklin Tree (*Franklinia Alatamaha*) was received as a gift from the Bartram Association of Florida, in commemoration of the 200th anniversary of the birth of William Bartram. The president of the association, Mrs. Millar Wilson of Jacksonville, presented it to Dr. Hamilton Holt and it was planted close to the Winter Park entrance, where it flowered profusely with large white flowers with yellow centers. The rarity of this tree, now extinct in the wild, stemmed from the fact that the seeds were not fertile and all Franklin Trees were derived from cuttings taken from the only native specimen known to botanists.

The opening day crowds had also unfortunately missed by a month or so a spectacular and unusual flowering display on the hillside garden close to the second mirror pool of the creek. Visitors that day had been shown the site that was to be part of a unique horticultural trial to try and answer the question of whether tulips could be successfully grown in Florida; the general opinion at the time being that they could not.

Edwin Grover was instrumental in setting up this experiment. In New York City over the summer of 1939, after attending the World's Fair, he read a newspaper clipping stating that after the fair closed for the season in October, a million tulips used throughout the site and given to the Fair by the Holland Bulb Growers with the aid of the Dutch Government would be discarded. The tulips were most prominently displayed at the Rose Court, a favorite spot at the World's Fair, where sculpted figures, apparently suspended in mid-air, and the statue *Builders of the Future* by the famous sculptor Zorach, were strikingly set off in a magnificent planting of linked hedges and thousands of tulips.

Grover wasted no time in writing to the New York headquarters of the Holland Bulb Industry, asking whether they might consider donating the bulbs to the Mead Botanical Garden rather than destroying them. He received a reply pointing out that the bloom the second year would only be about 65% of the first, and it would be better if new bulbs were planted. Nothing daunted, Grover suggested that in that case, they might be interested in providing ten-thousand fresh bulbs directly from Holland. This request was agreed to with the proviso that they would send different types and full instructions for planting, but would require a report at the end so they could judge the success of the experiment. They agreed to cold-store the bulbs in Holland and ship them to Florida before Christmas, to give a display bed which would bloom sometime in late February and early March. "If we succeed in bringing forth one-quarter of an acre of tulips in bloom it will be a unique attraction in Florida. It is going to take a lot of work to get the ground in shape, but I believe we can do it," reported Grover in a letter to a colleague.

When the bulbs arrived for testing in Florida, the gift amounted to over 16,000 bulbs, consisting mainly of Darwin hybrids, but with a few other types included. One thousand hyacinth bulbs were added as a bonus. All the bulbs came stored in wooden trays and, following the enclosed instructions, were transferred to the darkness of the Orlando Ice House for a period of seven weeks. By that time three-inch roots had developed and the first tip of the foliage had come through the surface. Then the tulips, still in the trays, were planted in full sun in the annual (hillside) garden. This planting was done on January 12; by February 20 the first Darwins were in bloom, and eventually the slope was covered with great masses of tulips in reds, yellows, purples, and whites. It was a spectacular display and attracted numerous visitors, many who found it an ideal subject for their new Kodachrome cameras.

For the purpose of comparison, a portion of the trays was not refrigerated but placed instead in white sand in a dark corner until roots and leaves had developed and then planted in moist, well-shaded peat soil. These tulips developed plenty of foliage but were mostly short-stemmed and reluctant to bud and bloom. In the center of the tulip garden, they planted the 1,000 Dutch hyacinths which had the same cold storage treatment as the tulips. By mid-February, these were beginning to show pink blossoms.

It was probably the first serious attempt to grow tulips in Florida, and the results showed that the Darwin hybrid tulips were more successful than other varieties and that cold winter storage was essential. The significantly reduced blooming for two-year-old bulbs was a further disadvantage so it was decided not to repeat the experiment the next year and the bulbs were grubbed up after flowering. The experiment confirmed the general opinion that tulips did not grow well in Florida.

2.2: *March 1940, and it's tulip time, the result of an experiment by Edwin Grover. The tulips were grown on the hillside garden which sloped down to the creek and the future site of the amphitheater.*

The tulip-growing experiment had been Grover's idea, but just at the time they were profusely flowering he suffered a hammer blow to his personal life with the devastating death of his only son, Graham, aged 25. Graham was a graduate of Winter Park High School and at one time a champion tennis player for the school. He later entered Rollins College, where he suffered a nervous breakdown, possibly a schizophreniform disorder. For the last two years, he had been working at the Garden and had made a rapid recovery, attributed mainly to the friendship he had struck up with Jack Connery, who was wisely guiding him in his work.

In the very early morning of Tuesday, March 5, around 2 a.m., C. M. Waite, engineer of the northbound Atlantic Coast Line passenger train on route to New

York, slowed his train as he approached the grade crossing at Minnesota Street in anticipation of stopping at the Winter Park station. Through his flickering headlight, he saw in the distance the figure of a young man standing by a pole at the crossing, who as the train reached the intersection suddenly fell forward and was hit by the engine. Waite applied the emergency brakes, brought the train to a halt, and rushed back to the spot only to find Graham Grover already dead. No inquest was held into the death, decided on by Peace Justice Eugene Duckworth who viewed the body at the scene, but suicide seems to have been the most likely reason for the incident. The sudden death of his son was the second tragedy in the life of Grover; several years earlier his wife had died of injuries after being struck by an automobile outside their home. Grover was by nature a very private person and we have no readout of how this calamity affected him, but it seems likely that he intensified his efforts in the Garden as some compensation for his loss.

Critically missing from the facilities of the Garden, with the thousands of visitors it was receiving, were bathrooms and simple places where people could sit down, partake of refreshments and purchase souvenirs to take home and postcards to send to friends. With the many flower shows and other events planned, time was of the essence to correct this, and the ground was broken in early February 1940 for the erection of a Reception Lodge and gift shop. The rustic building, finished both inside and out with pecky cypress to harmonize with the natural surroundings, would be 38 feet by 50 feet in size, located towards the center of the site, and would have public toilets in addition to the sales and reception area. Mr. and Mrs. Edmund Stowe of Sanford were appointed to be in charge of the shop, selecting items to sell and decorating the interior spaces, carrying out the theme of a rustic building in their decorations by using bamboo curtain rods. The first official postcards of the Garden arrived, all in full color, including scenes along the brookside trail and a particularly stunning card showing five orchid blooms. The Lodge, formally opened on March 17, 1940, also sold photographic equipment, color film, and soft drinks.

2.3: Left: Original wooden lodge, circa 1960. Right: One of the many postcards sold there.

The reception lodge became a natural display area for flower shows, social gatherings, and a space to showcase the work of painters and other artists who appreciated the natural and floral beauty of the Garden. One such artist was the eminent painter and illustrator, Harry Russell Ballinger, who arrived in Central Florida from New York City over the winter of 1939 and chose Orlando for his home. As one of the best-known illustrators in the country, famous for his black and white wash technique, he was under contract for twenty-five years with the *Saturday Evening Post* and the Hearst Publications to illustrate their leading stories.

In recent years he had shifted to painting, almost entirely in oils, and the move to Florida was driven by the desire to find new painting locations. Once he had visited Mead Botanical Garden, he declared it to be the loveliest place in this section of the South and returned nearly every day to paint. "The natural beauty is unspoiled. The colorful twisted tree trunks, the brook, the trails, the orchids, and everything just inspire me to go on and on painting," he stated. He ended up painting seven scenes: *The Sunlit Trail, At Lake Lillian, One Sunny Afternoon, The Sky-Blue Waters, Late Afternoon*, and two orchid paintings.

To coincide with the official opening Ballinger agreed that these paintings, together with three others painted in Central Florida, could be exhibited in the lodge. A reception took place from 3 p.m. to 5 p.m. to give the public an opportunity to meet the artist and inspect the lodge; the pictures remaining on show until March 24. The artist presented the painting *The Sunlit Trail* to the Garden, and it was gratefully accepted by Grover, who had it hung in the lodge.

In February 1941, the lodge became the scene for another artist reception, this time for the local and popular Sam Stoltz, known in Winter Park for his eccentric house building designs and nature-inspired murals, as well as his portrait work in oils. On the walls of the homes he built, Stoltz was fond of creating fantastically painted frescoes and decorative reliefs depicting birds, fish and other local wildlife. This time the subject matter for him was not the Garden itself but the man behind it.

How he came to choose Theodore Mead as a portrait subject is not known, but the inspiration for the painting design probably came from an existing photograph from 1927 of Mead in his greenhouse, surrounded by orchids. Stoltz's interpretation, using the grid method, was typically unconventional and different, giving Mead a likeness so real that many people felt he was just about to speak. In the painting, he is surrounded by images of the rare and beautiful orchids he developed, to give the picture the defining essence and spirit of his life.

Having completed the portrait, Stoltz donated it to members of the Garden Board, who were delighted to accept it and suggested, as a way of thanks, a gathering at the Reception Lodge beginning at 3 p.m. on February 2, 1941. It was open to the general public, with the artist bringing some of his paintings, tables, and screens, so there was a chance for everyone to have an opportunity to discuss Stoltz's unusual work with him. At the end of the event, the Stoltz portrait of Mead joined Ballinger's *The Sunlit Trail* on the walls of the Reception Lodge.

2.4: Top: In 1940, H. R. Ballinger, eminent illustrator of the Saturday Evening Post, chose the picturesque scenes of Mead Botanical Garden to paint in. Bottom left: A photograph of T. L. Mead dating from 1927 was used as the basis for an oil painting of him surrounded by orchids by Winter Park artist, Sam Stoltz (bottom right).

With the first touch of spring in 1941, the collection of 10,000 amaryllis bulbs grown by Mead were coming alive and sending up flower spikes. A few had started blooming and hundreds more were showing buds. The earliest Hemerocallis, or daylilies, were also beginning to flower and it would be just a few weeks before hundreds of Mead's hybrid daylilies would be hanging over Howell Creek, "reflecting their golden beauty in the waters of the brook."

2.5 *An exuberantly colored postcard of a portion of Howell Creek, probably based on the black and white photograph shown in 1.11, top left.*

By May and early June, the jeweled growth of the 30,000 Mead caladiums in the many shaded corners was beginning to show, as were the gardenias, including the patented Belmont, the Hadley, and the Veitchii varieties. These blooms were heavy with fragrance and had grown so well that they had become crowded and would require moving after flowering. Grover was fond of telling the story about how the Garden obtained one particular rare type of gardenia. According to Grover, Jack Connery was out looking for plants and saw an ancient and rare gardenia tree in a cow pasture near the St Johns River. He said to the farmer, "Say, mister, how much do you want for that tree out there in your pasture?" "Not a thing," replied the farmer, "The roots of the darn thing broke off half of my plow last Spring. Just fill in the hole after you take it out and I don't care what you do with it." Grover always ended the tale with, "And that's the way the Mead Botanical Garden got one of the rarest gardenias in this whole state."

The Garden was fortunate in having two greenhouses completed in time for the opening. The large Porter one by the Orlando entrance, filled with a thousand orchid plants, had an oil heating plant that was only completed after midnight on the day of the opening, ensuring that all the plants came through the cold weather without injury. As well as the orchids, there were other unusual plants there that were recommended to visitors, particularly those of them with Kodachrome cameras. Generally, by mid-summer, several large clumps of the Bird of Paradise plant (*Strelitzia reginae*) were in bloom. A native of South Africa, the plant got its common name from an Australian bird of astonishing beauty and its botanical name from the family surname of the wife of George III, Mecklenburg-Strelitz, hence *Strelitzia reginae* or Queen Strelitz.

Also of note, near the side door inside the main orchid house, was a container of an extremely rare lacy, feathery fern, very delicate in appearance with many long stems that touched the ground around the pot and extended in many directions. This was the Asian Walking Fern, *Asplenium ruprechtii*, imported by Theodore Mead more than thirty years ago and found in East Asia and the Fiji Islands. When the tip of a leaf touched the ground, new plantlets sprouted, creating a walking effect that Mead always said showed that they were seeking their way back to the Fiji Islands.

The smaller propagating greenhouse, near the Winter Park entrance, contained many plant seedlings and cuttings. In charge was Mr. Herbert Thomas, who worked at one time with Luther Burbank, and later with the Drier Seed Company. Experiments were underway treating cuttings with and without vitamin B as a root-promoting hormone.

2.6: Top: Dwarf evergreen azalea beds fronted the magnificent Porter greenhouse by the Orlando entrance. Bottom left: The Orlando gatehouse with formal annual flowering beds. Bottom right: The propagating greenhouse by the Winter Park trail entrance.

Each year new improvements were made to the Garden layout and design, new plant groupings established, and annual seedlings planted in previous years began flowering. As Grover put it, "We by no means offer this as a completed garden, but as a beauty center which will continue to grow and to include new beauties continually from now on." He suggested that the invitation to visitors could best be summed up with the slogan "Watch Us Grow."

In December 1940, the Waxahachie Nursery, Texas, had donated plants for a large formal rose garden. Earlier, Connery had mixed peat and topsoil and prepared a large raised bed on the right-hand side of the trail just inside the Orlando entrance, in anticipation of the roses arriving. The Orlando Garden Club agreed to provide fertilizer and sent a ton of dairy compost to be worked into the soil. When the shipment of 650 roses came, it included *Étoile de Hollande, Radiance, Kaiserin Auguste Viktoria, Mme. Pierre S. duPont, Editor McFarland, Maréchal Niel, Talisman,* and a dozen Cherokee roses in each color, pink, white, and red. Many were in bloom the following April.

The low-lying region around the Orlando gate was prone to flooding, due to spillover from Howell Creek or after heavy rains, so the formal annual garden next to the rose garden at the Orlando gate was also raised, and enlarged to nearly twice its original size. The Orange County Board of Commissioners helped alleviate the local waterlogging by laying 250 feet of drain tile, digging a drainage canal through the azalea garden and the caladium garden, and increasing the size of the intake at the curb near the entrance from Nottingham Street. Cherry laurels were planted at the corners of the formal garden in anticipation of the planting of thousands of annual bedding plants currently growing in the nursery area.

In late October 1944 a hurricane came through the Garden and felled a few trees and palms that blocked the trails until cleared, but thankfully left the greenhouses and other buildings untouched. One of the two tall sentinel Washingtonian palms at the Orlando entrance crashed through the fence, leaving the gatehouse marked by just one tall palm.

Jack and Helen Connery had both given several years to the development of the Garden and had suffered a significant financial sacrifice as a result. Now, with two small sons as part of their family, and a home on Richmond Road in Winter Park, Jack needed to supplement the small income he and Helen, as secretary, received from the Garden. Money was so tight that at times they were living a hand-to-mouth existence, and it is likely that this led to Helen's hospitalization for malnutrition. Dr. Gartley, who attended her, told Jack Connery that it would be unwise for either of them to return to working at the Garden, and he needed to get a "proper job." Luckily, an opportunity came along that would put more than just bread on the table.

In anticipation of being sucked into the hostilities in Europe, the Air Corps had commandeered Orlando Municipal Airport and reopened it as the Orlando Air Base in December, 1940, where it was to become a center for fighter plane training under the command of Colonel Voss. Initially, the tract was over 100 acres of a low-lying cow pasture near wasteland, and the job of transforming the landscape to an attractive and functioning air base was secured by Jack Connery. He got the job as landscape engineer based on his work at Mead Botanical Garden and particularly his expertise in planting palm trees. In March 1941, the giant landscaping project got underway under his supervision. Large quantities of rich loam were hauled into the base and spread to a four-inch thickness, hundreds of tall cabbage palms were dug up along the St Johns River, transported to the camp and transplanted, and thousands of potted plants, hedging material, and flower cuttings were acquired from all over Central Florida. In addition, thousands of smaller palmettoes were set out, areas grassed over, hedges planted, and paths mulched to complete the transformation, making the whole area appear from the air as if it were a regular part of the Central Florida landscape.

The final task was to camouflage the runways, and for this they required them to be spread with areas of peat to disguise their straight lines, and then observed from the air. Professional peat suppliers were contacted and gave different samples for a comparison test. Connery decided to dig out a quantity of peat from the Garden close to Howell Creek to compete with the suppliers. When the results were in, the Mead Garden peat was judged the best, to which Jack replied "You can have all you need if you will take it out the way we want it done."

So, in August 1942, and under Connery's supervision, the army dug out 7,500 cubic yards of peat, and in exchange the Garden acquired two large lakes of an irregular and natural shape, from 10 to 14 feet deep and 100 to 150 feet wide. Planting of the banks with palms and azaleas then followed. Howell Creek fed the southernmost lake, then the north lake via a channel, before that water flowed back into the creek, creating an island joined by three bridges. A further small island was formed in the south pool.

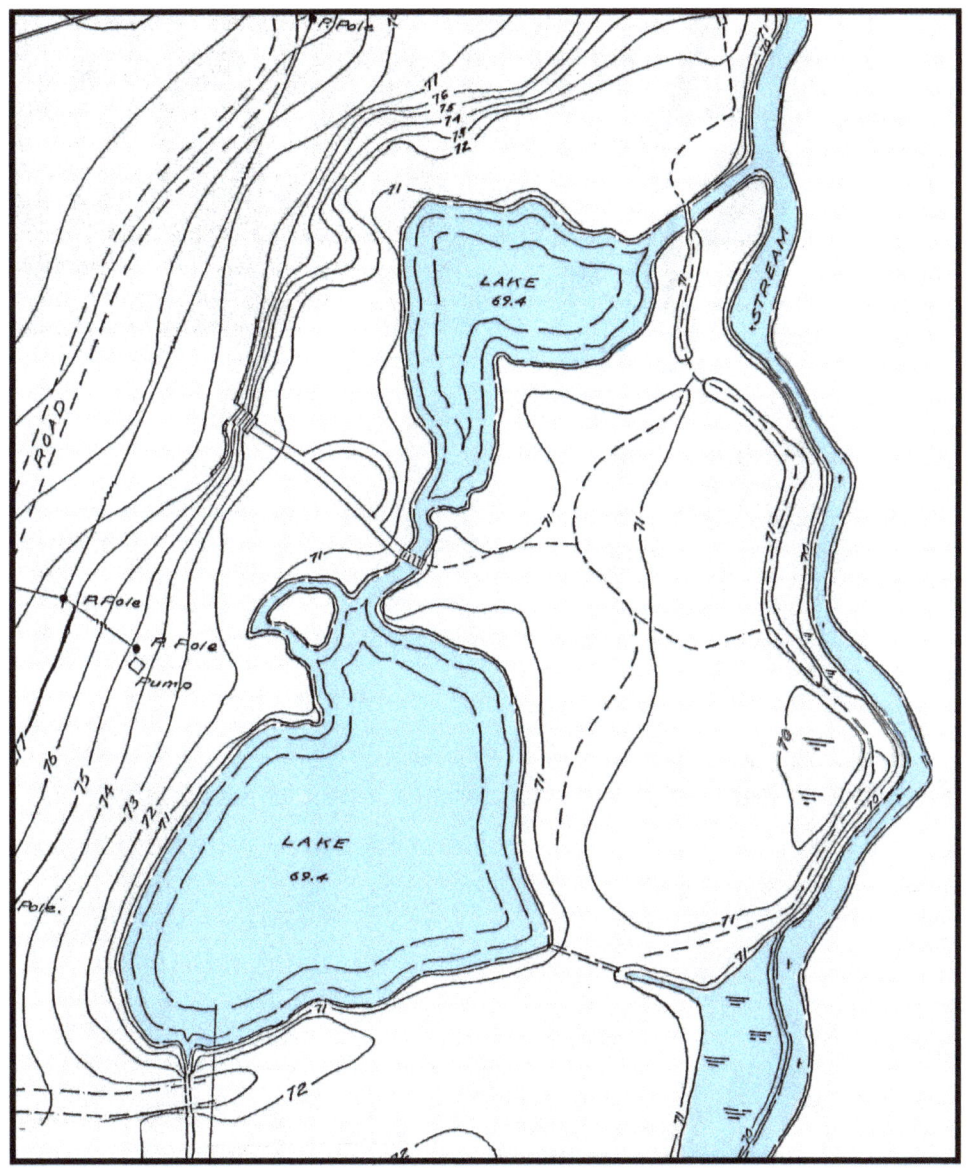

2.7: *In the summer of 1942, the part of the Garden just south of the reception lodge was extensively re-landscaped. There was a critical need for supplies of peat to camouflage the runways at the nearby Orlando airbase where Jack Connery was working. He agreed to provide the peat so long as it was taken out in a way that created an island and two large lily pools connected to Howell Creek.*

The Garden Develops Beautifully

The two large lakes transformed the layout of the Garden, introducing larger areas of water to balance out the tropical plantings and allowing the display of tropical water lilies, including the blue, pink, yellow and white varieties. In other parts of the Garden, thousands of calendulas, snapdragons, pansies, carnations, and larkspur annuals in the formal garden by the Orlando gate were in full bloom. The annuals in the hillside garden were matching their display with four large beds of scarlet salvia aflame with color and interspersed with accents of the sky-blue Chinese forget-me-nots.

In 1939, when Grover was in New York on Rollins business, he visited the annual flower show at the American Museum of Natural History and was thrilled by the magnificent spectacle. He was taken with the idea of running annual flower shows by plant-type at Mead Garden, and even one for the whole of Central Florida, with the Garden taking the lead. He wrote to Martin Andersen, "Why should not Central Florida put on an annual flower show in the Colosseum that would attract thousands of people from all parts of Florida. Both nurserymen and private growers would exhibit, and I believe an exhibition of national importance could be developed in the course of a few years. The Mead Botanical Garden would be glad to take the lead in organizing the exhibition and I see no reason why it could not be started in a modest way the coming winter. What do you think of the idea?" Andersen's reply is not recorded, but we can imagine that he would have been positive.

Smaller plant-specific flower shows were initially held throughout 1940, with such success that many became annual events. Mead was well-known as a pioneer in the breeding of hybrid amaryllis, receiving his original stock from Henry Nehrling, and thousands of his hybrids would be in bloom in the Garden by early April. The collection featured amaryllis with beautiful shadings from deep reds to light pinks, the first all-white hybrid amaryllis, and the classic striped Mead-strain amaryllis that were common in gardens all over the South throughout the 1920s and 1930s. Consequently, the American Amaryllis Society decided to hold their seventh annual Southeastern Regional Meeting at the Garden on April 13 and 14, 1940. Dr. H. Harold Hume, Dean of the College of Agriculture at Gainesville, was appointed the chief judge, assisted by a group of other amaryllis experts, and the exhibits were to be featured in the new rustic lodge near the annual hillside garden.

For a short period to coincide with the show, the Society loaned the Garden a rare lavender-blue amaryllis, *Worsleya procera*, which was exhibited in the main greenhouse. The sight of this incredibly rare plant in bloom was a delight to visitors and exhibitors alike. Sometimes called the Empress of Brazil, this exotic flower grows from a bulb to become about two feet high, the long bulb necks lifting the flowers into the air. The flower color varies, from almost white to deep lilac-blue, sometimes with spotted petals. It was claimed that the only other bloom in the entire United States was to be found in Miami.

2.8: Left: *Mead-strain amaryllis;* Middle: Worsleya procera, *the blue amaryllis of Brazil;* Right: Mead's Lemon daylily.

In early Spring, any visitor walking along the brookside trail could not fail to be enchanted by the 2,000 or so Mead's daylilies flowering alongside the creek. These were mixtures of the common species, mainly the lemon-yellow *Hemerocallis flava*, and the tawny-orange *Hemerocallis fulva*, and his named hybrids, 'Mead's Lemon' and 'Chrome Orange', the latter a clear orange evergreen variety considered one of the finest of the early bloomers. The American Amaryllis Society took the hemerocallis (daylily) under its wing early in its existence because of the natural close relationship with the rest of the amaryllis family, leading to the first and second National Daylily Shows in the United States being held under the Society's auspices at Mead Botanical Garden.

The first was held on May 18 and 19, 1940, with Ralph Wheeler managing the show with the cooperation of Connery and Grover. Exhibits were displayed in the lodge and included a special showing of varieties and breeding selections from The University of Florida's College of Agriculture, whose program focused on the breeding of new daylilies suitable for growing in Florida. The American Amaryllis Society show was repeated the following year on March 29 and 30, 1941, and was extended to include daylilies and other associated bulbous plants, with show categories for crinums, zephyranthes (rain lilies), hymenocallis (spider lilies), and alstroemerias (Peruvian lilies).

Ambition grew with the planning for an Inter-State camellia show, set for January 12 and 13, 1941. As the first big show of its kind, it would be open to amateur and professional growers and interested visitors from all the Southeast states and sponsored by the Garden and the garden clubs of Orlando and Winter Park. There would be show classes for all categories of flower blooms, arrangements, and corsages and the Orlando Art Association would be cooperating with a competition for artists featuring camellia paintings in any of three mediums; oil, crayon, and watercolor. All specimen flowers and collections would be shown in the main orchid house, and all arrangements and artists pictures in the reception lodge. As the show neared its opening, camellia blooms began pouring into the committee for judging.

Judging day was the day before the show opened and best flower in show received a silver dish, with bronze medals and blue, red and white ribbons awarded in every class, and cash prizes in the art division. On the day, the individual flowers were artistically mounted on panels of monk's cloth, each bloom hanging in its small corsage tube of water, creating an impressive display that covered both long walls of the orchid house. Throughout the Garden, committee members had organized for hostesses to be available to help direct exhibitors and members of the public. The newspaper reported attendance in the thousands, and commented that the show featured some of the most exquisite camellias one could ever hope to see; it was simply "quite a sensation," surprising even the customarily reserved Jack Connery.

2.9: Left: During flower shows, the reception lodge was used to display submitted blooms. Pictured is the May 1940 National Daylily Show exhibit. Note postcards for sale on wall behind. Right: Helen Connery admires six-feet-tall blue delphinium blooms at the Florida State Flower Show in May 1941.

Building on all this flower show success and experience, Grover saw no reason why they shouldn't go for the home run – a Florida State Flower Show – as he had suggested to Martin Andersen following his experiences when in New York in 1939. Of course, a much larger venue would be needed, and the Orlando Coliseum on the north shores of Lake Ivanhoe at 1640 N. Orange Avenue was chosen, with dates arranged for May 3–5, 1941.

Built in 1927, this large Moorish-style building was the center of social life in Orlando at that time, with a skating rink, dance floor and imposing stage that was the scene of many big-band concerts. It had a solid maple floor and 36-foot-high ceilings affording unobstructed views of the stage area. For the flower show, the arrangement reached was that the Garden Mart of Orlando would have a formal garden of flowers in the center of the enormous dance floor, and the fourteen spectator booths built around the dance floor would be used for unit exhibits. Exhibit booths were already reserved for the Mead Botanical Garden gift shop, the Florida State Parks, and the Winter Park Garden Club. The entire show would be under the sponsorship of the Mead Botanical Garden with Helen Connery as chairman and endorsed by The Florida Federation of Garden Clubs. All garden club members throughout Florida were asked to help in organizing and working on the show. Every flower grown in Florida, whether it be annuals, flowering shrubs or flowering trees, was eligible for entry, and the idea of a show received enthusiastic support from growers all over Florida because it would be the first show where amateur and professional growers and florists could come together in a spirit of cooperation.

Close to the opening day, a large banner announcing the event was hung in front of the Coliseum and cutting the ribbon to open the show fell to Mrs. Harry A. Griffin of Daytona Beach, President of the Florida Federation of Garden Clubs. Admission was 50¢ plus 5¢ tax and hundreds gathered for the show. Much talked about was the Garden Mart's 25-foot by 29-foot formal garden exhibit, surrounded by a three-foot brick wall, with a central plot of grass and potted bougainvillea trees at the corners, borders with Easter lilies, pansies, ageratum, dracaena, and pink impatiens contrasting with brighter red geraniums. Also attracting interest was a display of huge delphiniums from Joseph Leinhart's Oviedo Gardens, some of them with monster flower spikes four to six feet in height. They were placed in tall alabaster vases and ranged in color from white and pastel blue to deepest purple-blue. Overall the show was judged a great success, and this was considered due mainly to the hard work and organizational skill of Helen Connery, as chairman.

There were flower shows at the Garden throughout the rest of the 1940s. The second annual Inter-State Camellia Show was planned for January 18, 1942, and had an advance registration of more than double the previous year. This year there would be two orchid houses available to double the display space in which to exhibit the specimen blooms and collections, reserving the lodge for the flower arrangements. Once again, the show was jointly sponsored by the Garden and the garden clubs of Winter Park and Orlando. Camellia *Alba plana* was one of the Mead Garden collection that was singled out as one of the most long-lasting blooms for buttonhole or display.

2.10: Left: Camellia Alba plena, Right: Camellia Glen 40.

The show took place annually, generally around the second or third week of January depending on the weather, to increasing audiences and submissions of show blooms. In 1947, at the seventh annual show, the new No. 3 greenhouse was used in addition to the other greenhouses to display over 1,000 named camellia blooms to the delight of the 2,500 exhibitors and visitors. In 1949, the ninth show was sponsored by the Winter Park Garden Club. The 50¢ admission fee admitted flower lovers to the Garden, to the orchid house, and to the displays of camellia blooms in the greenhouse and arrangements in the lodge. The tenth annual show in 1950 was described by Robert Finfrock, president of the Mead Botanical Garden, as one of the most brilliant and sparkling in the long history of these garden exhibitions. Best Flower in the Show award went to a magnificent dark crimson 'Glen 40' camellia blossom entered by Mr. and Mrs. William A. Warrick of Winter Park.

The Garden's first camellia show was held in February 1940, shortly after its opening. Eight succeeding shows attracted more than 10,000 camellia lovers and growers, and did much to popularize this aristocrat of winter-blooming flowers throughout Central Florida. The Inter-State Camellia show had been a great success, but it was clear that with the growing interest in camellia culture Mead Garden was too small for Inter-State or even Florida-wide shows. Consequently in the 1950s the Florida State camellia show moved to the much larger Municipal Auditorium at 400 West Livingstone Street, leaving local plant societies the use of Mead Garden.

The eleventh annual Camellia Society of Central Florida event, sponsored by the Winter Park Garden Club, was held January 20 and 21, 1951, with the theme *Peace in Flowers*. Fresh sawdust on the trails and a new gate at the Nottingham entrance welcomed visitors. The admission price of 75¢ included a tour of the Garden and orchid house, viewing of the prize camellia blooms, and the presence of costumed hostesses directing guests to points of interest. Shows were held in 1952 and 1954, and the tradition continues today (the 73rd annual camellia show was held on January 19, 2019).

Daylilies came back to the Garden in late April 1945, with one of the largest displays of hybrids ever shown in the South and all the best varieties labeled by name. Daylily hybridizing was a Cinderella story that had the plant group in a no-man's land of little interest until the large-flowered, bright yellow hemerocallis 'Hyperion' was introduced in 1924. Interest continued to grow and Dr. Arlow B. Stout, of the New York Botanical Garden, an early pioneer in daylily hybridizing, started a breeding program based on seeds and species plants obtained from the Far East, which resulted in some 50,000 crosses and hundreds of thousands of seedlings. Stout selected and introduced approximately 100 cultivars, his first, 'Mikado', dated from 1929. After the end of World War Two the process took off, particularly in Florida, where the advantage was that the plant did not have a dormant period, so seedlings flowered in one season making it possible to discard unsatisfactory plants quickly. Through these combined efforts, nearly every color range became possible from the yellow, orange, and pale pink of the species, through vibrant reds, purples, lavenders, greenish tones, near-black, and near-white. It was a transformation of the daylily into new colors, with larger and more abundant blooms. The 1945 exhibit was dominated by the private collection of Ralph Wheeler of Winter Park, who was one of the leading Florida hybridizers, producing blooms of many colors including crimson, rose, scarlet and bi-colors. Wheeler was so skilled and prolific in his hybridizing that the International Bulb Society awarded him the Herbert Medal in 1947 for his outstanding work in advancing the knowledge of bulbous ornamental plants. Interestingly, previous holders of the Herbert medal had been Henry Nehrling and Theodore Mead; both awarded the medal posthumously in 1937.

The daylily show reappeared in 1949 and featured more than 50 of Wheeler's choice and named hybrids. Other exhibits were from Wyndham Hayward's Lakemont Gardens that included many of the crosses made by Stout, and one from Soper's Florida Gardens featuring the famous Russell hybrids. The annual daylily event, held on Sunday, May 19, 1950 was a special event. It featured Ralph Wheeler and Wyndham Hayward of Winter Park and Kenneth Soper and Mrs. G. H. Knight of Orlando as exhibitors, together with prominent daylily hybridizers Dr. John V. Watkins of the University of Florida, H. M. Russell, based in Spring, Texas, and Dr. Arlow B. Stout, of the New York Botanical Garden. The largest crowd in the 13-year history of the show occurred on April 29, 1951. This event featured more than 700 varieties of blooms, and was so popular that a second show was held on May 27 for later blooming varieties. Outstanding among the flowers on view at that time were some of Wheeler's named varieties, including his sensational purple-black 'Raven', the lovely 'Arla', the showy orange 'Naranja', the huge ruffled 'Show Girl', the wine-colored 'Bacchus' and 'Amherst', the orange-yellow 'Brassy Sun', and his pastel novelty 'Lilac Time'.

2.11: Left: Ralph Wheeler judging blooms at the Daylily Society Show at the Mead Botanical Garden in 1951. Top right: Wheeler's medal-winning 'Naranja'. Bottom right: Hybrids of a lavender coloration, like this one 'Lavender Flight', were prized as likely to eventually produce a true-blue daylily.

Wheeler and Hayward would be the driving force for continuing a daylily show at the Garden for many years to come, but the growth in interest in hemerocallis resulted in many regions setting up local societies and running their own shows supported by local hybridizers keen to sell their creations. Wheeler's daylily quest as a hybridizer was to create a true-blue daylily, and many of his hybrids in the 1950s featured large elements of lilac and lavender color. At the 1955 event, he showed the daylily 'Prodigy', a dusty rose color but the bluest of any daylily so far developed, and the one he was using in experimental pollination efforts to establish the first pure-blue daylily. Wheeler died in 1963 without achieving the true blue but with numerous awards to his credit, with hybrids such as 'Naranja' (1947), 'LilacTime' (1951), 'Playboy' (1954) and 'Nevermore' (1956). Many purple, lavender and part-blue daylilies have been produced since Wheeler's time, but the all-over, true-blue cultivar still eludes the hybridizer.

The first ever caladium show was also held in the Garden in July 1940, celebrating the 30,000 caladiums originally from the Mead Oviedo estate in full leaf in the shady nooks and crannies of the Garden. The first person and every twenty-fifth person were presented with a caladium plant of their choice from a wide range of varieties. Since caladium hybridizing had started in the early 19th century, there was by 1940 a bewildering proliferation of near-identical cultivars all with different names, a position that the daylily was rapidly approaching. Once the number of varieties of any plant group reached the thousands, how does a hybridizer create something of memorable character? Theodore Mead raised such a question in a 1915 letter he sent to Egbert Reasoner, relating to Henry Nehrling's hybrid caladiums, "They are beautiful of course but it is almost impossible to surpass previous introductions, and most of the new ones can be pretty well matched by older sorts, many of which have disappeared from cultivation as their successors are pushed." Even so one of Mead's hybrids has survived as the brilliant 'Blaze', renamed no doubt for marketing purposes from the original name of 'T. L. Mead' given to it by Nehrling.

In addition, there was a display of 40 amarcrinums, an intergeneric cross between an amaryllis and a crinum. The crosses, almost certainly between *Crinum powellii* and *Amaryllis belladonna*, had been made by botanists in Italy and also in California. The blooms were pink and funnel-shaped and sweetly scented like a tulip.

A feature of many of the flower shows was the gift of a boutonniere of colorful flowers to every lady visiting the Garden. The fragrance and sheer number of blooms on the gardenia bushes meant that this was a favorite item to give away in April/May time on Gardenia Days. In April 1953, a Gardenia Week was organized to celebrate the extensive plantings of gardenias in full bloom and recognize the gift from the Sanford Garden Club to buy the initial nucleus of the collection. Now, newly grown and healthy, the flowering shrubs made a white wall of fragrance along Gardenia Walk – forming, it was said, one of the largest concentrations of gardenia blooms in the state.

2.12: In the shady areas of the Garden, and down by Howell Creek, were sweeping clusters of caladiums with their vibrant foliage color combinations of white, pink, red and green. Well represented were Mead's hybrids such as 'T. L. Mead', now rebranded as 'Blaze' (left). In spring, the Gardenia Walk was heavy with fragrance from hundreds of the evergreen bushes (right).

In 1950, with the encouragement of the American Begonia Society, the Garden hosted the first Florida show devoted to the begonia. The show took place in the lodge April 15 and 16 and featured Begonia Rex hybrids and many more in the cane-stem, rhizomatous, semperflorens and tuberous groups. The first annual Central Florida Hibiscus Show was held at the Garden on Sunday, June 24, 1951.

The writing skills of Edwin Grover were used to full effect in the newspaper articles he created to publicize the Garden. As Grover wrote them, his friend Martin Andersen published them. Grover calculated that by the end of 1941, since opening, a total of ninety-two articles had been carried in the Orlando daily newspapers, with one nearly every Sunday, and forty-seven outside newspapers had carried a total of 94 of his press releases. Also, Grover had written a personal letter to every garden club president and every woman's club president in the State, inviting attendance and describing the attractiveness of the Garden. Despite all this local publicity, after the initial novelty had worn-off, residents appeared relatively indifferent to the splendor of plants, flowers, and orchids from all parts of the world. Martin Andersen likened it to the prophet in his own land whose people fail to appreciate the beauty to be found there.

2.13: Part of the rose garden near the Orlando entrance, which had a display of more than six hundred and fifty roses.

This indifference did not deter Grover, who was fond of expansively describing the experience of a tour through the Garden. One example, written in 1941 before the two large lily pools had been excavated, was titled "Beauty Spots in the Mead Garden – A Leisurely Walk along its Wooded Trails." In the article he described the horticultural riches of the Garden starting at the Orlando entrance with its large formal rose garden of six hundred and fifty roses, and then leading to the Orchid House fronted by a formal planting of annuals. From there a trail followed the winding brook of Howell Creek with its three mirror pools and many shadowed vistas, part-lined with golden daylilies, originally from the Mead estate. Another beauty spot he pointed out was the "Treasure Island Pool," where there was a small azalea and caladium covered island with a large cypress tree at its center. Farther along the trail was a caladium garden of more than ten thousand bulbs of Mead's hybrids, and the gardenia garden. The Reception Lodge and gift shop were located about halfway between the two gate-houses, and next to them was a large hillside garden of annuals at the foot of which was a second attractive mirror pool. Turning to the left led to the camellia garden and beyond that the trail wound through another section of jungle growth, past another pool and tiny waterfall to a large display of azaleas near the Winter Park entrance. From here, a short trail led to Lake Lillian, which was hidden in a dense jungle growth.

The description ended up back at the Orchid House, which for many visitors was the outstanding feature, containing thousands of different orchids of about fifty varieties. When the Garden first opened it was the only public exhibition of orchids in Florida, and for many visitors it was the first time they had seen such an expansive display.

CHAPTER 3

The Orchids of Mead Botanical Garden

An orchid enthusiast visiting Mead Botanical Garden through the Orlando gate in the early 1940s would have walked past the sweeping 250-foot wall of red, white and purple sweet peas, through the entrance, and then would have been immediately drawn to the left and the imposing Porter greenhouse, clearly marked "Orchid House." Inside, Mead's collection of around 1,000 rare orchids would have been an amazing sight even to an expert. Our visitor would have seen masses of orchids of the genus *Cattleya*, including the species *C. amethystoglossa, C. skinneri, C. maxima, C. forbesii, C. velutina, C. granulosa, C. schroederae, C. trianae, C. warscewiczii, C. dowiana, C. bowringiana*, and *C. lawrenceana*, while hundreds of never-seen-before hybrids would have crowded out the display shelves and hanging rails of the greenhouse. If the expert knew much about orchids and the rarities he gazed at, the estimated value of ten thousand dollars for Mead's collection would have seemed a credible figure. Many of them were rare species, or one-off primary hybrids that at that time could easily sell for hundreds of dollars each at an orchid auction.

3.1: *The largest (Porter) greenhouse was close to the Orlando entrance, bordered in the early years with red, white and purple sweet peas (top). Palms, azalea bushes, and a bird bath framed the entrance of the greenhouse, bottom picture dating from 1948, which also shows, left to right, William Jess, female visitor, and one of the gardeners.*

Had our imaginary visitor been a non-expert, the resident orchidist at hand in the greenhouse would have helped by pointing out some of the corsage orchids, with their common names – the Christmas orchid (*C. trianae*), the Queen orchid (*C. dowiana*), and the Easter orchid (*C. mossiae*). He would have continued by showing visitors Mead's hybrid between *C. lueddemanniana* and *C.* (*Laelia*) *tenebrosa*, which Mead first planted as a seed in 1897. He named this cross "ludbrosa," following a naming system that he invented, but his pet name for it was his "Star Boarder" because it just sat in his greenhouse and looked at him for seventeen years before it decided to reward him with a flower.

The brief tour might then have taken in the various Dancing Lady orchids (Oncidium alliance), Moth orchids (Phalaenopsis alliance), and Slipper orchids (Paphiopedilum alliance). The Coconut orchid (*Maxillaria tenuifolia*), named because of its coconut scent, and the Dove orchid (*Peristeria elata*), the national flower of Panama, would have been other orchids of interest, as would the commercial Vanilla orchid (*Vanilla planifolia*). In one corner of the greenhouse, a group of terrestrial Nun's orchids (*Phaius tankervilleae*) would have intrigued any visitors had they investigated the hooded bloom to discover the part of the flower that looks like a praying nun. Finally, if the timing were right, the beautiful rare blue orchid of Burma (*Vanda coerulea*) might provide a final jaw-dropping flower to gaze at and complete this brief exploration of the world of orchids.

The original Mead orchid collection size had multiplied with donations, exchanges with other botanical organizations and growers, and the development of mature flowering orchids from smaller rooted offshoots. Shipments arrived of 75 orchids from Baldwin and Company, which included a stunning example in bloom of *C.* Queen Mary; two cases from Brazil containing 27 orchids from the Parque Indigena (Park for Indigenous Plants) of Santos; and an extensive collection from the Panama Canal Zone. This group contained more than sixty plants of eleven different varieties, eight of which were new to the Garden. Orchids came in from the swamplands and forests of Columbia, courtesy of a Mrs. Austin, and from Tom Sawyer, a young orchid collector based in Winter Park who spent much of the year collecting orchids in the jungles of Venezuela. Initially, he sent a case of *C. mossiae* which when divided and potted made 130 individual plants. At a later date, he shipped quantities of *C. speciosissima (lueddemanniana)*, which added around 150 plants to the collection once potted up.

3.2: The greenhouses were stuffed full of orchids with at any one time hundreds in bloom, as these original photographs demonstrate.

3.3: A selection of some of the Cattleya *orchids on display. Clockwise from top left:* Cattleya maxima; Cattleya tenebrosa; Cattleya dowiana; Cattleya trianae.

Sometimes there were so many different varieties of orchids in bloom that the orchid experience was a bit like being part of a world tour. There were clusters of white pansy orchids with red centers, native to Columbia (*Miltoniopsis*); long graceful sprays of orange-yellow orchids from Brazil (*Oncidium auricula*) that waved with the slightest breeze and bloomed continuously for over three months; several lovely white orchids with deep golden lips from the mountains of India, China and Southeast Asia, known simply as Thunias (*Thunia marshalliana*); a spider orchid from Mexico (*Brassia caudata*); an explosion of red orchids from Costa Rica (*Cattleya Guatemalensis*); a yellow corn-cob orchid found in New Guinea and the Philippine Islands (*Robiquetia cerina*), and the breathtakingly beautiful Butterfly orchid (*Psychopsis papilio*) from Trinidad, Central and South America, said to have been the orchid that triggered the Victorian mania for orchid collecting in Europe in the 19th century.

Many of the Cattleya-alliance hybrid orchids had exquisite blooms, but one was singled out by the *Orlando Sentinel* above all others in November 1940. Registered in 1931 to Mr. George Baldwin as *Cattleya* J. Palliser, it was a large and fragrant orchid of bright rich purple color, having a pronounced purple lip shading into a golden throat. The seed parent was *C.* Rosalind, a child of *C. dormaniana* and *C. trianae*, and the pollen parent *C.* Saint Gothard, a child of *C.* Gottoiana and *C.* Hardyana. It was just about possible in 1931 to trace the original species of *Cattleya* and *Laelia* used to create this hybrid as; 25% *C. dowiana*, 25% *C. trianae*, 12.5% *C. warneri*, 12.5% *C. warscewiczii*, 12.5% *L. tenebrosa*, and 12.5% *L. purpurata*. This single rare orchid could easily have commanded $500 at auction.

According to Jack Connery, visitors to the greenhouse often asked if there were any green varieties among the many orchids. If the season were right, they would be shown a collection of *Cypripedium* orchids in bloom that included the green slipper orchid (*Paphiopedilum Maudiae*) from Asia. These orchids stand up on tall, slender stems growing out of mottled green leaves, often with just a single bloom, and have a distinctive cup-shaped lip, hence their common name of Venus or Lady Slipper orchid. Connery was of the view that many men preferred their exotic beauty to that of the frilly feminine *Cattleya*s, describing them as "real men's orchids."

3.4: Many other orchid genera were also on display. Clockwise from top left: Vanda coerulea; Laelia purpurata *(now* Cattleya purpurata*);* Cypripedium chloroneurum *(now* C. chloroneura*);* Oncidium papilio majus *(now* Psychopsis papilio*).*

Florida native orchids were also well represented in the orchid greenhouses. There were examples of Florida's most unusual leafless orchid, *Dendrophylax lindenii*, the Ghost orchid that flowered between June and August. On a trip down to the Everglades, Connery had found samples growing in large cypress hammocks, partially submerged in water, and brought several logs with orchids attached back to Winter Park. On account of their rarity, these proved to be powerful items for exchange with other botanical institutions and William A. Frederick, the superintendent of the Washington Botanic Garden, supplied them with a large selection of mature plants as a swap for a four-foot log carrying a display of the ghost orchid.

3.5: *Florida native orchids were also featured in the orchid collection, including the enigmatic ghost orchid,* Dendrophylax lindenii *(left); the stunning bee swarm orchid,* Cyrtopodium punctatum *(middle); and the delicate Florida butterfly orchid,* Encyclia tampensis *(right).*

Several examples of the Cigar orchid (*Cyrtopodium punctatum*), sometimes called the Bee-Swarm orchid, were regular bloomers in the orchid house, creating a massive display sometimes over a meter across and individual inflorescences capable of producing hundreds of one-inch brown and yellow flowers. Examples of the Florida Butterfly orchid (*Encyclia tampensis*) were also on display, and the collections on view represented an era when native orchids were a common sight particularly in South Florida before development and indiscriminate wholesale collection from the wild drove them to the verge of extinction.*

A highly unusual and educational part of the orchid display, at one end of the greenhouse, was the incubator where there was a progressive display of the germination and early growth stages of baby orchids. This consisted of a number of air-tight flasks, containing baby orchids barely visible to the naked eye, lying on a layer of agar jelly mixed with salts and nutrient sugars. Other flasks contained little orchids at various stages of their development. This display came from Dr. Lewis Knudson of Cornell University, whom Mead worked with in the 1920s, and was the co-discoverer of this technique of orchid seed germination in sterile laboratory conditions. After growing for a year or two, the little orchid plants would be one-quarter to one-half inch in size and would be ready to be transplanted into small pots. After anything from seven to ten years, they would have grown to flowering sizes. Sterile conditions were necessary at all times, since if fungus spores gained entrance to the flask they would multiply and destroy the baby orchids.

In the 1940s this display would have been interesting to a visitor but it would be several decades before its real relevance would be appreciated. The Knudsen/Mead method allowed thousands of seedlings to be raised to maturity from a single orchid seed-capsule, producing large quantities of plants economically. It was the scientific breakthrough that made commercial orchid growing possible, and together with tissue culture, ensured orchid affordability to the public and launched the orchid industry as we know it today.

By late 1941, the main Orchid House was becoming badly overcrowded with around 2,000 orchids and other rare plants, but help was at hand. A friend of the Mead Garden, on a visit to the greenhouse, suggested the need for a second

---

* On a more positive note, today the Knudsen/Mead technology of sterile micropropagation is being used by the Fairchild Tropical Botanical Garden to reintroduce millions of young native orchid plants into South Florida's urban landscapes where they once grew in abundance.

display house and generously offered to purchase and erect it without cost to the Garden. A 14-foot by 27-foot model was chosen from Garden Mart on N. Orange Avenue and the ground was broken in December 1941. The plan was to devote the present Orchid House 1 exclusively to the growing and exhibition of orchids, particularly those that needed a warm house, while Orchid House 2 would be used to house *Cypripedium* and other orchids that required a cooler house, as well as rare tropical plants of other varieties in the garden collection. Relocated to the new orchid house would be the large collection of *Cymbidium* orchids, commonly known as Boat orchids, that were sent by L. Sherman Adams, one of the largest growers of orchids in New England.

A third greenhouse followed in 1946 and occupied a site on the south shore of the southern of the two large lily pools, and in 1947 money was given for a further small greenhouse for propagation purposes. This brought the total number of greenhouses to five, which included the small greenhouse by the Winter Park entrance.

In 1943, the Garden received a visit from the American Orchid Society as part of their Visit your Neighbor series, and subsequently a report of their visit appeared in the house magazine. They reported how the Garden included a mile of wooded trails, five large pools, two of them with small islands, a slat house filled with sub-tropical plants and three greenhouses, where the orchids seemed to be in a healthy, excellent growing condition, with several plants of Brassolaeliocattleya Niagara, *C. Cassadaga*, and *C. forbesii* being in bloom. Other species noted in flower were *Catasetum macrocarpum*, *Oncidium altissimum*, *Dendrobium Phalaenopsis*, *Laelia anceps*, *Vanda tricolor*, and *Epidendrum cochleatum*. In addition, they noted three native orchids growing in the garden area, namely, *Stenorrhynchus orchioides*, *Epidendrum conopseum*, and *Habenaria strictissima* var. *odontopetala*. The article concluded, "It is a fitting and lasting tribute to a man who contributed much to Florida horticulture that a memorial to him can provide a place for those who desire to get really close to Nature."

3.6: *In addition to the large Porter greenhouse by the Orlando entrance, and the small propagating greenhouse by the Winter Park trail entrance, there were three other greenhouses in the Garden.*

3.7: Grover stands in the main greenhouse (left) by a flowering Vanda tricolor *(right).*

In the 1890s, when Mead started to hybridize orchids, his initial starting point was the genus *Cattleya*, and what could be achieved with interspecies crossing. Transferring pollen between species was a skillful task and one where timing was critical, especially if the two orchids were in flower at different times, when the pollen had to be carefully stored. Following pollination, the seeds then had to be harvested when ripe and viable. As a seed parent receiving the pollen, Mead found that *C. bowringiana* (now *Guarianthe bowringiana*) gave him more hybrids brought to blooming size than any other species. Among the many primary hybrids he created were crosses between *C. schilleriana* and *C. maxima* and between *C. bowringiana* and *C. forbesii*. In 1904, these two were registered by the International Orchid Register of London, respectively, as *C*. Meadii for the man and *C*. Oviedo for his residence.

He was also intrigued by the possibilities of intergeneric crossing of *Cattleya* with the genera *Laelia* and *Epidendrum*. In addition, he also wanted to cross an *Epidendrum* with *Laelia*, to create Epilaelias. One of his early crosses between these two genera involved crossing *Epidendrum fragrans*, with a creamy white shell-like flower with purple lines radially across its hollowed spoon-shaped lip, and *Laelia flava*, with purplish flask-shaped bulbs and bright chrome-yellow flowers on a long stalk. He called this hybrid Epilaelia flagrans, which he admitted was a sort of Alice in Wonderland portmanteau word as it combined the names of the parents. This hybrid would certainly have been on display in the greenhouse, among the many hundreds of others hybridized by Mead.

Naming new hybrids was a serious matter for Theodore Mead; he just couldn't subscribe to the fashionable practice of using the name of the hybridizer, as Luther Burbank did *ad infinitum*, or the hybridizer's wife, or relative, or acquaintance. To him, it made the most sense if the new name conferred something of the hybrid's parentage. So, he invented a new naming system that was designed to recognize hybrids without too great a strain on the memory. The idea came following the creation of several successful Epicattleya (now Guaricyclia) hybrids using *C. bowringiana*, and consisted of representing each species by one syllable or part of a syllable of its name and combining them, adding a vowel if needed for euphony. Thus bowringiana, he suggested, could be represented by boa, and among primary hybrids the result would be Epc. adboa with *E. advena*, Epc. plicaboa with *E. plicatum*, and so on. Mead's hybridizing efforts are recognized by eleven different named hybrids in the International Orchid Register. Using *Guarianthe bowringiana*, he is credited with Crismoloboa, 1894; Plicaboa, Adboa, Melodboa and Lodilboa, all 1900; Leobroboa, 1901; Amneboa, 1902; and Meadii, 1904. In addition, he is the named hybridizer of Cuco, 1902; Oviedo, 1904; and Ludbrosa, 2014. There must have been dozens more hybrids that he never got around to formally registering.

All these orchids would have been on show in the main greenhouse at Mead Botanical Garden, and to view the many unique Mead hybrids would have been an astonishing experience to even the most experienced orchid specialist. Woven into any storytelling about the *Cattleya* class of orchids that visitors would have heard in the greenhouse would have been the current importance in fashionable circles of power dressing with an orchid corsage.

3.8: Two Cattleyas *hybridized and credited to T. L. Mead;* Cattleya Meadii *(left) and* Cattleya Oviedo *(right).*

Cattleya cut-flower mania through the media of the corsage was rife during the 1940s and 1950s, dominated in the early days by species such as *C. labiata* starting off the season in the autumn months, followed by *C. percivaliana* over Christmas, then *C. trianae* into the New Year, and *C. mossiae* heralding springtime. These large-flowering *Cattleya* orchids were rare and exotic and established themselves as the classic flowers for glamour and sophistication to be worn as a corsage by high-status ladies and, most famously, by the wives of various US Presidents. Size really did matter and since the average *Cattleya* had flowers 5–6 inches across, to really draw attention to the corsage frequently more than one bloom would be used. Perhaps the greatest exponent of the power corsage was Mamie Eisenhower, who was rarely seen in public during her husband's two terms in office without her corsage of two or three *Cattleya* flowers.

Cattleya orchids were also a popular gift on hospital visits and as a corsage a symbol of prestige, often presented to someone, or even some organization, that was loved and admired. *C. labiata* in particular, with its frilled petal edges, was a favored orchid for presentations of this kind. It was the main type of orchid that Theodore Mead had grown and hybridized, and was on display in the main greenhouse. It was entirely appropriate then that after a few weeks of the opening of Mead Botanical Garden a distinguished female visitor to Orlando became a recipient of this token of admiration.

3.9: In the 1940s and 50s, the Cattleya orchid was an essential part of the orchid corsage for the rich and famous. Left: Mamie Eisenhower in Richmond, Virginia in the 1950s. Right: Jack Connery gives Gladys Swarthout, New York Metropolitan Opera Company's mezzo-soprano, an orchid from Mead Botanical Garden as she arrives in Orlando for a concert performance, February 1940.

Gladys Swarthout, mezzo-soprano of the Metropolitan Opera Company, was due to give a concert in Orlando on Thursday evening February 15, 1940 at the American Legion auditorium, located at 1024 N. Orange Avenue. As she stepped down from the train on her arrival in Orlando, Jack Connery presented her with a *Cattleya* orchid corsage from Mead Garden. Apparently, the operatic and concert stage star was in an ebullient mood upon her arrival and immediately pinned the stunning blossom upon her chic coat. At the concert, she sang and thrilled an audience of 1,300, performing a program chiefly consisting of folk songs, under the sponsorship of the Orlando-Winter Park Civic Music Association.

Many years later, in 1954, the female recipient was Lilian Gish, noted stage and screen actress; the donor Edwin Grover, who presented her a *Cattleya* named in her honor. The occasion was the attendance of Lillian at the Animated Magazine to give a presentation on "Art in the Theater" and the subsequent award of an honorary Doctor of Fine Arts degree by Rollins College.

By 1948, the Garden had grown steadily since its opening in January 1940 until it now had eleven permanent structures – five greenhouses, two attractive gatehouses, built of split, cabbage palm logs, a slat house for partial shade, a reception lodge with a living apartment for the superintendent, two small caretakers' cabins, and a storehouse.

3.10: William Jess, orchidist at Mead Botanical Garden, shows a visitor one of the many orchids on display, 1948.

Recently, Milo Shattuck of New York City had been appointed as the superintendent of the Garden. He was a graduate of Harvard University, studied horticulture at the New York State Agricultural College, and had experience working in the New York Botanical Garden. The orchidist of the garden was William Jess, who had been there since the mid-1940s. He was president of the Central Florida Orchid Society and vice-president of the National Orchid Society, and greeted guests in the orchid houses to describe the plants and answer their questions. At that time, in the late 1940s, the working staff of the Garden consisted of four men and two women who worked part-time.

Grover's love and knowledge of orchids developed with the Garden, and by the end of the 1940s he was regularly giving talks on orchids and how to grow them. His knowledge of native and ornamental plants was equally broad, so much so that Taylor Briggs, the first director of Parks and Recreation for the City of Winter Park, believed he had a degree in horticulture. Grover took a particular interest in the orchid incubator in the main greenhouse, where seeds germinated in glass vials and small plants grew by the warmth of a miniature electric bulb.

The decade had been a successful one in many ways, and the Garden had grown into a major attraction for tourists, at least during the winter months. Over 20,000 garden folders were printed and distributed throughout the State and the Garden was listed in guidebooks and in copies of Florida Tourist attractions. In the Greater Orlando area, it was one of only two locations worthy of a tourist's time according to the Florida State Advertising Commission, the other being Sanlando Springs. In 1949, the commission printed a map of Florida in a campaign to encourage Floridians to vacation within the State, and to recommend interesting sites for winter visitors.

Orchids were the star attraction for winter visitors and local residents alike, but as the decade wore on, orchids became more affordable and many stores started displaying and selling them at reasonable prices. Local societies sprang up to meet the interest in orchid collecting and growing. There was a trend, too, towards more regional flower shows that were in direct competition with the bigger annual shows in the Garden. As a result, there was an overall decline in attendance over the decade, signaling the start of financial problems that would doggedly stick with the Garden for the rest of its life.

3.11: Top: Grover checks on the plants in one of the smaller greenhouses, circa 1940s. Bottom left: Grover holds a Cattleya orchid ready for repotting outside the main orchid house, 1948. Bottom right: Grover inspects the orchid blooms in the main greehouse, 1953.

CHAPTER 4

Financial Difficulties and the City Takes Over

Edwin Grover's love of green spaces and Nature was hard-wired into his upbringing as a country boy in New England, where he roamed the local woods and fields and later spent five summers exploring Bible Hill, north-east of St Johnsbury, Vermont. In *The Country Boy's Creed*, written when he was at Rollins, he voiced the opinion that "life out-of-doors and in touch with the earth is the natural life of man." In Mead Botanical Garden he found a special place where he could turn these words into practice.

He believed strongly in the preservation of natural beauty and was in accord with the comment of Irving Bacheller, "one thing for growing cities to remember, with their spreading pavements, was the need to keep in touch with Mother Earth." Grover's warnings did not go so far as to predict specific population growth figures for Orange County. Instead he liked to refer people to the allegorical story concerning the groundbreaking ceremony of the Illinois and Michigan Canal in Chicago on July 4, back in 1836, where one of the speakers, from the top of a hogshead cask, grew so bold as to predict that someday the village of Chicago would have a population of 10,000 people. According to the story, the applause was so tremendous that the speaker was encouraged to increase his prediction to 100,000, but this was too much for the crowd, and they booed and jeered him off the barrel.

In Grover's view, with increasing population growth at a rate few could accurately predict, there would be a growing need for a greater number of peaceful, quiet and tree-shaded areas where flowers bloomed and birds sang. He was a passionate and vocal advocate of the need for city and county-wide funding to achieve this, using the example of Chicago and Cook County, who years ago had provided funds for the purchase of wooded areas and places of natural beauty. It was his fond hope that one day the agencies of the City of Orlando, the City of Winter Park, and Orange County might use part of their budgets to promote and develop Mead Garden.

As Grover looked at the annual attendance and income records for the Garden throughout the 1940s and compared it to expenditure, he became more and more worried. It had always been the intention that the maintenance of the Garden would be taken care of by charging a small admission fee, as was done at the Cypress Gardens and elsewhere in Florida. Initially, the entrance fee in 1940 was 25¢, then 35¢, then 50¢ in the later years of the 1940s, but even this did not appear sufficient. The problem boiled down to a lack of summer visitors who ranged in numbers from about two hundred to four hundred a month, compared to thousands during the tourist season. By comparison, typical monthly expenses in the early 1940s, with gasoline for the truck and electricity for lighting, were around $400 a month. It was clear that without winter visitors paying to see orchid displays and spectacular flower shows, the Garden could not cover its monthly salary bill, let alone develop a surplus for new project work. Despite the claim of record-breaking attendances, there simply wasn't enough local interest outside of the tourist season to keep the Garden solvent unless some new outdoor events could be created or developed.

Grover needed to put his thinking cap on for ways to increase the income-producing status of the Garden, particularly if they were ever to acquire the rich, peaty lands to the east of Howell Creek and create a proper entrance for Winter Park residents. Central to his thinking was the potential land asset associated with the twenty acres of the Bartels tract, including the claypits, which were not currently an essential part of the Garden. In 1945, right out of the blue, came one of the strangest proposals to use this land; for the relocation and reconstruction of William Randolph Hearst's 12th-century Cistercian monastery that could then become an integral part of the area's attractions.

MEAD BOTANICAL GARDEN

Paid Admissions: (50¢)

Month	1947	1948	1949	1950	1951
January		1082	1308	1001	
February		2213 93	2011	1735	
(Camellia Show)		(2062)	(1253)	(1115)	
March		1447	1628	1106	
April		816	784	749	
May		300	374	315	
June		300	263	181	
July		243	286	201	
August		351	271	182	
September		290	179	109	
Totals for the entire year:	10,636	10,632	9,723	7,224	3,464

4.1: Attendance numbers for the Garden for the period 1947 to 1951 when admission was 50¢.

In 1925, Hearst, while in the village of Sacramenia in the province of Segovia, Spain, saw an ancient monastery he liked and, acting on impulse, decided to buy it and its associated buildings. He paid $500,000 for it, built a new monastery for the current occupying monks, and brought in architects to supervise the dismantling. Every stone was given a number to correspond with a place on a reassembly chart and crated for shipment back to the USA in more than 10,000 packing cases. Built in 1141 by Alfonso VII, King of Castile, the monastery consisted of the cloisters, refectory, cells, chapter house and kitchen covering a plot 130 by 120 feet, and when reassembled, destined for Hearst's San Simeon estate in California, would be by far the oldest structure in the United States.

However, soon after the shipment arrived in New York, Hearst ran into financial problems that forced most of his art collection to be sold at auction, and the massive crates were stored in a warehouse in Brooklyn. One of the most monumental white elephants in art history was eventually sold at a subsequent auction for $19,000, with the purchasers stipulating that they wished it to be erected somewhere in Florida. There were many interested parties, not all from Florida, but most balked at the transportation and reassembly charges – a museum in Boston, for example, figured it would need $100,000 in addition to the purchase price.

4.2: *In 1945, Grover was keen to acquire William Randolph Hearst's ancient Spanish Monastery for the Mead Botanical Garden site, currently languishing in over ten thousand wooden crates at a Brooklyn warehouse (top). Funding was not forthcoming, however, and it was left to a couple of Miami entrepreneurs to buy the relic and reassemble it in northern Miami, where it now resides as the Ancient Spanish Monastery tourist attraction (bottom).*

Undaunted by this financial Everest, the Florida State Bureau of Publicity negotiated the relocation of the monastery to Florida and invited interested communities to bid for the ancient relic. Grover was all for the idea of adding the monastery to the art and educational landscape of Winter Park and agreed to donate the Bartels tract as a site. A new organization was set up, the Winter Park Hearst Monastery Association, made up of Hugh McKean, Grover, Harry Schenck, manager of the Alabama Hotel, Wyndham Hayward, city editor of the *Orlando Reporter-Star*, and Robert Finfrock, to study the feasibility of bringing the building to Winter Park and how it might be funded. Dr. Hamilton Holt of Rollins College generously offered to give an afternoon of his time with a trowel laying mortar between the stones of the Hearst Monastery if and when acquired for Winter Park, in the event of a continuation of the current labor shortage.

On September 19, 1945, the Winter Park City Commissioners voted to grant the required twenty acres for the location of the monastery. The Florida State Bureau of Publicity agreed to assume full responsibility for raising funds for transportation and construction but failed to raise any part of the money necessary to bring it to Florida. The crates continued to languish in storage until bought by two Miami entrepreneurs who reassembled them in northern Miami, where they became a tourist attraction known as the Ancient Spanish Monastery. In 1954, *Time* magazine called the reassembly of the 135,000 stones the biggest jigsaw puzzle in history.

After this setback, Grover wrote to Holt asking whether Rollins would be interested in taking over the Garden, stating that although the City of Winter Park owned the land, it had not taken an active part in the promotion nor operation of the Garden, other than to contribute $250 a year. Holt's reply has not been found but it is likely that the College's financial position at the time would prevent such an acquisition. Subsequently in 1949, Grover had to make a specific request to the City for $500, pointing out the increased operational costs. Without some form of the public funding that Grover had dreamed of and with rising costs and declining attendances, particularly over the summer, the writing was on the wall as far as the Garden's future was concerned.

When public funding came in 1951, it came with strings attached that Grover didn't like the look of. Initially it sounded promising, as the City of Winter Park agreed to take over the Garden and committed a grant of $7,500 toward its maintenance for 1952. On October 23, 1951, the City revealed its true thinking with plans for a city recreation center on the Garden site, making use principally of the twenty acres of the Bartels tract. Mayor W. H. McCaully promoted a plan which called for a swimming pool, youth center building, football field, two baseball diamonds, two volleyball courts, and four tennis courts. To assuage Grover, Robert Finfrock, then President of the Garden, stressed that the horticultural part of the gardens would be continued with plants and shrubs and a bird sanctuary providing features of interest to garden lovers and wildlife enthusiasts, and further augmented with a library for horticultural books and exhibits, an experimental laboratory, and an amphitheater. McCaully said that the project would cost around $50,000 and that it could be funded either by passing the hat or by a two-mill hike in taxation over three years, which he favored. McCaully continued to urge the development of the new recreation center at various meetings, receiving largely favorable responses, including one where the results of a survey among student teenagers in the community were discussed. By then, brick barbecue pits and tables, a dance floor with tables and chairs, and an auditorium had been added to the growing list of facilities, and the likely project cost had grown to $85,000. The student reaction was lukewarm, as a newspaper report at the time concluded, "lack of public transportation dampened the reception of the proposals."

The City was also bullish when it came to its immediate plans for the Garden, with the City Manager Earle Harpole reporting that trees had been cleared off the extension of Garden Drive as far as the site of the proposed amphitheater, and indicated that this road would eventually connect with the garden entrance on Pennsylvania Avenue. He also promised that Winter Park residents would have a new parking area, located just across the street from the Winter Park entrance to the Garden, on the land donated by Mr. and Mrs. R. F. Leedy, and that that would be ready in just a few weeks, by late March, 1952. Neither of these two things happened.

Now in charge, the City stepped up the pressure via a series of uncompromising financial demands of humiliating magnitude and disdain for the founders of the Garden. It refused to release any more money for the Garden unless all the principal creditors who advanced money to the Garden, including all the community-minded citizens who bought revenue certificates, waived their claims as a total loss, and also that the Garden turn over all gate receipts to the City. This was a real poke in the eye for the Garden, and the dignified and upright college professor from Maine did something he probably didn't do very often, and snapped.

Incensed, he took up his pen and wrote an open letter to the *Winter Park Herald* entitled "So That the People May Know" that was published on June 12, 1952. In it, he castigated the City for suggesting that those residents of Winter Park who created the Garden should lose all their money as a reward for the public service which brought to the City property valued at more than $100,000. He lambasted the idea of handing over all gate receipts, emphasizing that it would deprive the Garden of its only source of earned income. He underscored this last insult as being "equivalent to a death sentence." Finally, he reminded them of the deed conditions under which the City had agreed to accept title to the land and keep and maintain it as a botanical garden, and criticized the City for failing to maintain the Garden properly. He pointed out that the land would have reverted long ago to its original owners had it not been for the Mead Botanical Garden Inc., and the few private citizens who gave their services and their money to preserve the valuable property for the city.

But the City had them over a barrel and they knew it. Ignoring all of Grover's protestations, the commissioners simply leased the property to Mead Botanical Garden Inc. in July 1952 for another year at $1, with Robert Finfrock as president, and waited for the inevitable. To make sure it was clear to everyone, the city restated that in leasing the property they assumed no obligation for debts or other items.

By December 1952, the inevitable capitulation occurred. Mead Botanical Garden Inc. was reorganized, electing former Winter Park Mayor McCaully as president, and, most importantly, agreeing to cancel the $23,000 debt owed to its founders. Helen Dunn-Rankin, immediate past president of the Winter Park Garden Club, was elected first vice president; Ralph Wheeler, prominent horticulturist, second vice president; Helen Connery, who served as the first secretary of the garden, secretary; and Raymond Greene, Mayor of Winter Park, treasurer. In announcing their release of all monies owed to them, Grover and Connery expressed the hope that the Garden would now be in a position to realize its goal of becoming a major Central Florida tourist attraction and botanical center.

4.3: *It is December 1952, and seated at a table an acquiescent Jack Connery (left) and a grim-faced Edwin Grover, co-founders of the Mead Botanical Garden, are shown as they sign waivers of all money advanced by them during the early days of the Garden's development, equivalent to $23,000. Mrs. Eugene R. Shippen, a trustee, and Robert Finfrock, immediate past president of the Garden, are shown witnessing and accepting the waivers, as control of the Garden passes to the City of Winter Park.*

It wasn't long before other agencies eyed up the Bartels tract and were all for trying to develop it. In mid-1953, the Orange County Board of Education focused its attention on the land as a possible site for a new Junior High School for Winter Park, offering $26,000 for it in a formal bid to the City, roughly half its appraised value. The City commissioners turned down the offer but, being desperate for money to maintain the Garden, offered to let the school board buy as many of the twenty acres as it wanted at $3,000 an acre. This put Grover in a bit of a bind – he realized the City needed money to maintain and improve the Garden, and this development ought to provide that – but was worried that a busy school next to a botanical garden might not be the best combination, nor fit in with the deed restrictions of the original landowners.

He elaborated these thoughts in a newspaper article where he questioned the wisdom of selling off an area for a school site when the City of Winter Park, with its present rapid growth, would need more parkland in the next 25 years. And if it was to be, should there not be a referendum on the decision? In the event, the sale stalled because of the inability to compromise on price. The *Orlando Sentinel* was sympathetic to the plight of both parties, and posed the question "Which would you prefer, sell part of the land, or lose the Garden entirely?"

In the background, Grover had his own compromise plan that could potentially deliver money for the maintenance, improvement and augmentation of the Garden as a botanical resource at the price of a limited area of housing on the Bartels tract. It was a variant of the pitch he'd made to Hamilton Holt in 1946 suggesting Rollins take over the Garden. This time he floated the idea, in a meeting between the Mead Botanical Garden Board, the City of Winter Park, and Rollins College, of the college taking a 99-year lease on the property, and dividing a thirteen-acre section of the Bartels tract into 40 city lots with an approximately assessed value of $2,000 each. This would generate $80,000 that could be used in part for the construction of an amphitheater on the old claypit site for Rollins College-produced plays, Animated Magazine events, symphony concerts, and professional productions during the winter season. A percentage of the proceeds from the theatrical productions would be dedicated to Mead Garden annually for its maintenance.

This Grover-led initiative died a quiet death on the grounds it was too massive an undertaking for Rollins and also by a rejection from college authorities of year-round productions in the claypit. In the meantime the Orange County authorities, having investigated numerous sites for a Winter Park school, including within the City of Orlando, came back to the Bartels tract as their favored option. They threatened to build in Orlando as the only other option and this galvanized opinion in favor of the Winter Park site. At a public hearing meeting on October 13, 1953, members of the Mead Botanical Garden Association pleaded for its use either for garden purposes, recreational area, or as a site for an amphitheater for outdoor productions, and indicated that present policies were to keep the Garden intact. Despite these pleadings, the majority of parents present voted for the Mead Garden site for the school. The City Attorney, Webber Haines, was given the task of clarifying the legal status of such a move, in light of the Bartels deed restrictions.

This public vote was as good as a green light to the Orange County Superintendent of Schools, Judson Walker, and an improved offer was made of $27,500 for 13 acres of the Bartels tract, which included a proviso whereby the city would be asked to make the claypit available for suitable use by Winter Park schools. The stated policy of keeping the Garden intact was in shreds and on November 19 four new directors were appointed to the Mead Botanical Garden board, most likely as a result of resignations.

In light of the strength of feeling from parents in the community, the county supported the addition of a referendum on the matter to be added to the regular ballot at the city's general election on December 8. The question for voters was "Do you favor the use of the undeveloped portion of Mead Botanical Garden as a public-school location?" The *Orlando Sentinel's* position was clear with the headline "Schools Try Grab of Mead Garden," which went on to question the sanity of sacrificing a third of the land for school purposes, and the legality of the Winter Park electorate deciding on a matter that many citizens of Orlando and Orange County had financed.

The board of directors of the Mead Botanical Garden were opposed to what they interpreted as an attempt by the school board to acquire the property through a political maneuver in getting the question in a public referendum. In their view, if the school faction in favor of the sale were successful, this would nip in the bud any chance for the garden's future as a botanical park and would also mark a betrayal of a sacred trust by the city.

The 84-year-old Edwin Grover roused himself from his peaceful retirement and went into battle one more time for the Garden. To anyone who would listen, and specifically to rejuvenate Garden supporters, he reiterated his arguments and those of the board. Of course, Winter Park needed a new local school, but this wasn't the site for it. A thousand young people and an area of rare and valuable plants just didn't go together – already young vandals had entered the property, trampled plants and broken into the garden lodge where they ransacked papers and photographs. Then there was the appraised value that he'd been told was $94,750 for the development as a subdivision versus the $27,500 the County was offering. Finally, there was the wording of the Bartels deed as a gift specifically for botanical use. He painted a bright picture of the Garden's future with C. D. Walker, president of the Walker Fertilizer Co., as a new director and under consideration to succeed McCaully as president of the garden. He mentioned that Dr. V. R. Gardner, former head of the department of horticulture at Michigan State, had agreed to act as a director of the garden; and that the Board was already in contact with a young man with a doctor's degree in horticulture for the job of superintendent. He summed up by describing the Board's vision for the future of the Garden: continuation of the collection and development of all types of tropical and sub-tropical plants; further improvement of the acreage so that at the earliest possible time it could become the actual botanical park that was initially conceived; and construction of a living botanical laboratory for the study activities of students in local schools and colleges.

The *Orlando Sentinel* of December 3, 1953, featured an editorial with a cartoon of Grover defending the Garden against school board intrusion and the "So the People May Know" headline. The subheading read, "Mead Garden Belongs to All Orange County and Shouldn't Be Cut Up for School." The article stated "The *Sentinel* is not convinced either that the schools need this particular site, or that the surrender of even so small a part of so valuable a development is in the public interest. There is plenty of other land in Winter Park which could be used as a school site." It continued, "It is short-sighted policy to say that because the particular twenty acres desired by the schools are at present undeveloped, the garden would not miss them. It is not large enough with all its 55 acres and full development has been delayed only by lack of funds. These in time will come and the importance of Mead Botanical Garden to Central Florida and the entire state will be more fully appreciated as its development becomes complete."

4.4: In 1953, the Orange County School Board attempted to acquire the 20 acres of the Bartels tract for a new Junior High School for Winter Park. The City of Winter Park was in favor of the plan but, with Grover leading a spirited defense, the idea was soundly defeated in a referendum. This cartoon appeared in the Orlando Sentinel who supported Grover's side of the argument.

The text further declared that the newspaper never normally meddled in other town's politics, but on this occasion, because so many Orlando and Orange County residents gave their money, time and material for the Garden, they felt that the residents were owed an accounting – moral and legal – should the Winter Park school trustees condemn and acquire this site for a school. The editorial finished with, "The *Sentinel* cannot escape the conclusion that the school board should look elsewhere for its site."

Voting took place on December 8, 1953, with the usual last-minute dire warnings from both sides. School board officials said that repeated surveys had shown no other site as suitable as this one; garden supporters that sale of the land would result in a serious legal challenge due to the deed restrictions. Voting was brisk, but when the polls closed and the votes were counted, the proposal to build a Junior High School on the Mead Botanical Garden site was rejected by a three to one majority, with 923 votes against the sale and 365 votes for it.

Having promised to take over the Garden in 1951, the city in 1955 moved to legitimize the situation following a request from the Mead Botanical Garden board for its cooperation in developing and maintaining the Garden as a beauty spot. City Attorney Webber Haines was asked to study the legal status of the garden under the city's trusteeship, to examine the operational costs of the Garden provided by the board, and to report back at the next meeting. On March 23, 1955, the city moved to take over operation of the Garden, with a positive operational balance of $204.71, and an operational running cost from 1954 of $5,428.90. Commissioners indicated that the botanical components of the property would be kept but that future plans for use would include a recreation area and community park. The deed transferring ownership from the Theodore L. Mead Botanical Garden to the City of Winter Park was signed on May 24, 1955. To execute their program, the City set up a park board as a standing committee in charge of the maintenance and development of all public parks in the city. This arrangement effectively dashed the hope of getting C. D.

Walker as the new president of the Garden, with his robust vision of its potential exciting future.

The City of Winter Park felt it should own the entire garden area, so it instigated the annexation of the two or so acres of land from the original Rose deed situated in the City of Orlando. The annexation question became part of a referendum on September 20, 1955, and was passed by voters.

In the early 1940s, Jack and Helen Connery were forced to leave the Garden to find employment that would produce a living wage, although Jack was still loosely connected with the Garden as "Director on Leave." Following his work for the Airbase in Orlando, the family moved in 1943 to Mount Dora, where he grew Irish potatoes, carrots, beets and beans on the rich muck land of Zellwood directly to the north of Lake Apopka, an area that was fast becoming the new center for vegetable production. On one occasion, he took a sack of potatoes into the Sentinel offices and handed them out for people to try. It was there that he admitted to having been given the nickname of "Corncob Jack" by the rest of the Zellwood farmers.

At Zellwood, he came across Dr. Brown Landone, an early leader of the New Thought Movement and Winter Park resident, who had an experimental farm in Zellwood growing large quantities of ramie, a source of natural strong fibers suitable for use in textiles. As a result of this meeting, in 1945 the family made a move to Lake Worth, where Jack was employed as Plantation Superintendent managing 1,500 acres of ramie for the Florida Ramie Products Corporation, under lease from the State Prison Farm No. 2 near Belle Glade. Edwin Grover visited the Connerys and toured the facility, describing the experience to Hamilton Holt in a letter and enclosing a piece of ramie fiber which he dared Holt to try to break. In September 1947, a hurricane destroyed the factory and plantation, which led to the family moving to Cuba where Jack secured the job as an agricultural engineer for Jose Aleman, a Cuban state official, who was looking to grow ramie on half a dozen farms. Unfortunately, after a couple of years, Aleman died and the Connerys returned to Miami, where Jack worked for a landscaping company and Helen took up secretarial work.

By 1951, they had moved back to Central Florida and settled in DeLand, where Jack's brother Joe lived, with Jack securing the post of landscape engineer for a major project in the area – the renovation and transformation of the springs at DeLeon, just north of DeLand, into a Florida water-themed roadside attraction. In 1925, the local residents had altered the original name of Spring Garden to the Ponce de Leon Springs, claiming this connection with the Spanish conquistador in common with many other locations in Florida and hence, logically, that the springs there were the site of the mythological "Fountain of Youth." The project that Jack Connery became a part of was to create tropical gardens and a landscape that could accommodate water tours, a jungle cruise and even a water-skiing elephant. The property became one of over a hundred roadside attractions in the state when it opened as Ponce de Leon Springs in May 1953.

Jack and Helen Connery established a plant and landscaping business in DeLand – Mimi's Nursery and Landscaping Company. The name Mimi came from Helen's childhood nickname given to her when she was a little girl with a habit of going around saying, "Me, Me." The nursery started growing a range of foliage and flowering plants – azaleas, camellias, poinsettias, and rex begonias, and the gesneriads, gloxinia and the African violet. Helen became President of the DeLand Garden Club and both the Connerys gave regular talks on horticulture at various garden club meetings. After a few years they decided to specialize in the hybridization and cultivation of African violets, turning this into a successful business. This took off with the landing of an exclusive deal with the F. W. Woolworth Company to supply violets to all their stores in the south-eastern USA. The business grew to the extent that they were supplying thousands of plants weekly to Woolworth's. In addition, they were contracted to give plant clinics and display their new varieties at Woolworth stores in exhibits entitled "Pathways of Beauty" and "A Vacationers Dream." To attend the clinics, they bought a station wagon and fitted it with shelves to carry the violets, making sure that on their travels they included plenty of locations where they could indulge their passion for fishing.

4.5: Following their departure from the Garden, Jack and Helen Connery eventually settled in DeLand, where they established a thriving nursery business, Mimi's Nursery & Landscape, growing a range of plants (top). They eventually specialized in the cultivation of African violets and their largest commercial customer was F. W. Woolworth (bottom left). They visited the stores to conduct plant clinics in support of the sales effort, making sure they managed to leave enough time to indulge their joint passion for fishing (bottom right).

Since leaving the Garden, the Connerys had kept in touch with developments there and were named as Garden Board members for the years 1952 and 1953. Even though he had signed his rights away to all the plants he received from Theodore Mead, Jack was still concerned about them, and particularly the condition of the orchids. He had heard that many were in need of repotting and some had become infected with scale. Over the summer of 1953 he organized for Edwin, his youngest son and at that point a junior at DeLand High School, to spend time working in the greenhouses alongside David Clarence McConnell, affectionately known as Mac, who was in charge of the orchids. Edwin recalls that period he spent tending the orchids, but his most persistent memory was the monotony of the lunch menu – Mac was a vegetarian and every day for lunch they had white bread sandwiches spread with peanut butter and honey.

Throughout that period and beyond, each year there would be calls for volunteers to help clean-up the Garden and take care of the plants, including the orchids, under the supervision of garden superintendents. Volunteers, and once the Garden became free in 1960, locals and sometimes visitors, were in the habit of taking any plant they took a fancy to and Edwin recalls this as a real problem. He remembers his father summing it up with the comment that "many a person's orchid collection was started with orchids from Mead Garden."

There are no records to tell us how Jack Connery saw the city takeover and the forced handing-over of all his rare plants, but he appears to have accepted it philosophically. He believed his creation would always be operated as a botanical garden. The idea that any part of the Garden might be turned into a recreation center with baseball diamonds would have horrified him, and one can only guess at his reaction had he known that in twenty or so years all of Mead's collection of plants would have vanished. But in 1955 he was confident that he could run the Garden and make it profitable, and made a formal request to the city for a 50-year lease on the garden property, with 5% of all gate receipts going to the city. To support his case, he pointed out that it was he who contributed the twenty-thousand-dollars' worth of plants to set up the Garden in the first place. When the City refused this request, he informed the city attorney Webber Haines in May 1955 that he would seek legal redress unless a satisfactory arrangement was worked out. Without the deep financial pockets needed to bring a claim like this, no more was heard of it.

Jack Connery kept in touch with Edwin Grover and moved on with his life. He was deeply disappointed at the way things had turned out but felt that he had done his bit and fulfilled his promise to Theodore Mead. Now it was up to others to carry the standard forward.

CHAPTER 5

The Amphitheater and Fashions in the Garden

It wasn't all doom and gloom in the 1950s for the Garden but, needless to say, lack of funds influenced and drove just about every decision. There was a chronic shortage of money even for basic maintenance in the late 1940s and there was a need for events that were income-producing and brought people into the Garden. Board member Dr. Eugene Shippen called on an advisory panel of about twenty influential members to come up with possible fund-raising ideas, out of which emerged Helen Dunn-Rankin's suggestion of a fashion show with the beautiful garden forming a natural background for the production. The proposal met with enthusiastic approval, and Fashions in the Garden was born.

The idea took hold and the sloping site near the lodge that once displayed magnificent tulips, but now was an untidy hay-field, was selected as the place for an amphitheater to stage the event. With a working committee and ably assisted by the loyal members of the Winter Park Garden Club, Helen Dunn-Rankin, supported by Elizabeth Shippen, set about clearing and beautifying the area. William McCaully designed a wooden stage at the foot of the slope and, with an eye to economy, built it to a scale so that it could be dismantled and converted into tables and seating once there was money for a more substantial stage. There were no funds for decoration other than the natural offerings of the Garden. But an arch was borrowed from a friend's house, some azaleas from a nursery to flank the stage, and scarlet bougainvillea tied into nearby palm trees to form a colorful floral background for the exotic gowns and furs in the show. Also, a special parking lot to accommodate 300 automobiles was cleared behind the amphitheater, reached from the Nottingham Street entrance. The fashion show took place on February 25, 1950, with the theme *Around the Clock*. With limited seating capacity, guests were advised to arrive early for the 3 p.m. start so that they might occupy one of the many umbrella tables. Entrance tickets to the event were $1.25.

5.1: *Newspaper advertisement for Fashions in the Garden, 1952*

Both Winter Park and Orlando dress shops participated, and models displayed Spring and Summer fashions in chronological order throughout an imaginary day. Styles entitled *Wake Up You Sleepy Head* were followed by those appropriate to *Let's Go Shopping and Play Golf*, then by themes, *Luncheon at the Country Club*, *Fun at the Beach*, *It's a Very Important Dinner* and finishing with *Enchantment Closes a Lovely Day*. Dorothy Lockhart Smith was the narrator and the gala day included tours of the Garden, door prizes, music during the promenade and tea hour by Edmund Cushing on the Hammond organ, and flower girls giving out Mead Garden orchid boutonnieres.

This first fund-raising idea brought the Mead Garden Association $1,200, and its success convinced the advisory committee to continue the custom but, both crucially and wisely, they decided to insist that any profits would be used for improvements to the amphitheater and beautification of the Garden, and not for its regular maintenance.

Despite the early success, and for reasons unknown, Fashions in the Garden was not performed in 1951, but was resurrected in 1952 and held on Saturday, March 22. The theme, with Helen Dunn-Rankin as narrator, was *Family Portraits*, consisting of fashion collections in the categories, *Young and Old*, *Teen-age Chic*, *Fashions for Florida*, and *Duplicate Models*. There was a flower arrangement competition, garden tours, and background music during the show provided by Charles Civiletti, popular organist from the Candlelight Restaurant. The Leo Sunny trio from the Orange Court Hotel's fashionable Patio Supper Club provided the entr'acte feature, and The Orlando Little Theatre presented a skit as a part of the fashion parade. An Audubon exhibit created with the assistance of the Florida Audubon Society and materials borrowed from the Baker Museum at Rollins College attracted much attention. The event was judged a great success and raised about $1,000.

Fashions in the Garden was getting established as a regular event and a Winter Park tradition that people looked forward to, generally following the same format each year but with a different theme. The third annual edition of the show took place on March 14, 1953 with *Preview of Spring* as the theme. The program was opened by a marching band from the Winter Park High School, starting at the Nottingham Street entrance and finishing at the amphitheater. Jonathan Dunn-Rankin, Helen's son, was the narrator, presaging his future as a well-known actor and broadcaster. It was the turn of *Hearts and Flowers* in 1954, where an estimated 3,000 attended at a $1.50 ticket price.

In 1955, with the uncertainty of the city-take over in everyone's mind, the event again did not take place, compounded by the attempt by city commissioners to have the $2,935.43 proceeds of previous events sitting in the bank transferred to them for immediate maintenance use. At a subsequent meeting, the Garden Club dug its heels in and agreed to release the money for specific Garden improvements either to the amphitheater or the two entrances, both badly in need of beautification. In the end, an agreement was reached that some of the money would be used for a permanent cement and wrought iron stage for the amphitheater, constructed over the winter of 1955–1956 so it would be ready for the 1956 event.

5.2: *The original 1955 concrete stage with iron grille work (left), had permanent seating and lighting by 1959 (right).*

The original wooden stage, which cost $180, was converted into tables to be used throughout the Garden. John Armstrong designed the new arena, free of charge as a gift to the Garden; $1,500 was paid Shadix Watson Inc., Winter Park builders, for cement and labor, the firm having made no profit on the construction; and $1,000 was paid R. G. Coffman Co. to cover the cost, without profit, of iron grille work. Fashions in the Garden contributed $800 to the city for this work and declared that profits from future events would be used for seating, followed in coming years by a contribution toward permanent theatrical lighting and sound for the stage. The fifth Fashions in the Garden show, with the theme *Easter Parade in Paris*, took place on March 10, 1956, and raised a profit of $947.03, which went towards the cost of installing permanent bench seating.

The 1957 event turned out to be "The most beautiful show we've ever had," according to Mrs. Dunn-Rankin. With a theme of *Holiday in Rome* and a theme tune of "Three Coins in a Fountain," it was held on March 2 and featured Neapolitan music throughout the performance. The stage decorations included a running fountain at the center of the stage and Corinthian columns bearing out the show's theme. People remarked that the leaf patterns on the trees behind the stage formed a perfect background for the spring garments. There was a surprise opening to the show when Jonathan Dunn-Rankin, making his third appearance as narrator, rode onto the stage on an Italian-made motor scooter wearing a white suit, sunglasses and a black beret worn at a jaunty angle over one eye, Italian style.

In 1958, the theme was *Magnolia Blossoms*, with Edith Tadd Little as set designer and the Winter Park Concert Orchestra furnishing the music during the refreshment break. Profits of $553.90 were handed to the City of Winter Park to help pay for the installation of lights, which came in the form of a donation to the Fashions in the Garden committee by the Florida Power Corporation when they installed new fluorescent units on Park Avenue and had no need of the old lamp posts. Edith Tadd Little and Helen Dunn-Rankin persuaded the city to donate the old lamps to the amphitheater, and they were set around the area, electrically wired with funds from the fashion shows, and hung with vine-planted baskets in Spanish style.

The eighth annual production on March 21, 1959, celebrated *Orange Blossom Time*, and featured Emily Bavar, Florida Magazine editor, as fashion commentator supported by Jonathan Dunn-Rankin's narration. Bavar arrived at the opening of the event in a decorative cart, dressed in a Gone-with-the-Wind-period antebellum costume, and was gallantly handed down from the carriage by Dunn-Rankin, also in uniform as a Southern gentleman.

Over the summer of 1959, plans were put forward for the construction of a dressing room building that would be ready for the 1960 event. Helen Dunn-Rankin presented the City with a check for $508, together with a promise of an additional $1,000, as contributions to the estimated $3,000 cost. She argued, "Since Fashions in the Garden assumed the financial responsibility for creating this place of beauty from what was a horrible jumble at no cost to the city except labor, we feel the city might match our funds for the dressing room." The City's initial and usual response of "not in the budget" met Mrs. Dunn-Rankin's indefatigable temperament head-on and there would only be one winner. The city commission awarded a sum of up to $2,500 toward the project, taking this amount out of the budget for re-landscaping Central Park.

5.3: *The amphitheater today with the dressing room flat roof visible to the back and right of the main stage.*

The proposed dressing room building would be 26 feet by 20 feet and contain separate dressing rooms with lavatory facilities for men and women. The plan was to locate it at the south-east side of the stage, low down and behind a large tree so that it would be partially hidden from the view of the audience. Construction of the dressing rooms was done at cost by local contractor Allen Trovillion.

The fully equipped amphitheater could now be used, with the usual rental fee, for all kinds of other cultural events, both during the day and in the evenings; and it was decided to form "The FIG Society" to manage and organize the use of the facility. Helen Dunn-Rankin was named the president and Edith Tadd Little the chairman in charge of productions. On October 1, 1959, the amphitheater was formally dedicated in an evening reception and program attended by mayors from neighboring cities, officials from the Orlando and McCoy Air Force Bases, city members and local dignitaries. Over one hundred local performers took place in an hour-long show featuring two ballets, two choruses, an orchestra, and many outstanding solo performances. The 1959 Community Christmas Sing was held in the amphitheater and it would become a regular site for concert recitals in the future.

On April 9, 1960, the ninth annual Fashions in the Garden event took place with the theme *Aloha to Hawaii*. In an attempt to maximize income, patrons buying tickets in advance were asked to donate $10 to FIG, which would entitle them to have their names printed in the program and receive two free tickets. As part of the event, hostesses and flower girls would be in native Hawaiian costume, booths would be decorated in Hawaiian themes, and the tea table would feature a bountiful display of Hawaiian fruits and pineapples. On the day, Hawaiian gowns from Shaheen's of Honolulu and twelve local shops had displays, with stalls on the mall featuring accessories for spring and summer. During the intermission a series of Samoan dances with ukulele and guitar accompaniments was performed. Once again it was rated "best ever," and judging by the number of attendees (600) and the profit generated ($2,200 compared to a previous high of about $1,500), this was indeed the case.

5.4: *The theme for Fashions in the Garden was 'Orange Blossom Time' in 1959 (top), and 'A Galaxy of Fashions' in 1962 (bottom).*

Home Town USA was the theme in 1961, featuring the activities of the American woman, and the clothes she wore – from early morning coffee in the garden, through attendance at church, including sports, the cocktail hour, tea, and evenings in the cool of the night. The event played to a capacity crowd, with all 750 seats occupied.

A nation obsessed with space, and the declaration that America would put a man on the moon by the end of the 1960s, strongly influenced the 1962 theme, *A Galaxy of Fashions*, and the stage decorations. Somehow, they managed to borrow a twelve-foot high model of a rocket from the Martin Company, which they staged with a large periscope, a metal mobile sphere, and a metal sun. Accessory and other shops were aligned along "The Milky Way" and intermission entertainment provided by the Orange Blossom Chorus. "Keep looking up" must have been the theme for the organizers as rain forced a rescheduling of the event to April 14.

5.5: In 1963, the stage was converted to look like a cruise ship (left), and colorful fashions were modelled by Rollins College students (right), echoed the theme of a 'Caribbean Cruise'.

The romance of a cruise captured the imagination of many, and dreams of moonlight nights on the Caribbean, cruising on blue waters, and the delight of shipboard friendships was recaptured during the twelfth annual show, *Caribbean Cruise*, set for March 30, 1963. The stage was skillfully converted to produce the impression of a cruise ship by George Rackensperger, and Tom Sawyer and

Edith Royal were shipboard commentators. Tickets were $1.50, and the event described as a "triumph."

As fate would have it, this would turn out to be the last Fashions in the Garden for some time. There was talk about starting a Fall version of FIG, but nothing came of it. Instead, someone came up with the idea of converting the event into a family fun day and calling it "Families in the Garden." A date of May 9, 1964, was selected and a number of family competitions planned, such as a hootenanny contest, an arts festival, and a square-dancing event on the Garden Club patio. Other attractions were a fish pond, mystery grab bags, a basketball toss, horseshoe pitching, strong man bell-ringing, a driving range, a cake walk game, a nail driving contest, pony and horseback riding, and ukulele strumming.

On the day, the Garden was open for family entertainment with picnic grounds and cooking facilities, and the usually quiet and peaceful atmosphere of the Garden was rent asunder with hootenanny music, bells clanging, horseshoes ringing, ukuleles strumming, children screaming and horses blowing. The classy and sophisticated Fashions in the Garden that returned thousands of dollars over the years to support the Garden was a thing of the past, and the goose that laid the golden eggs had turned into a scruffy and noisy barnyard cockerel. Compared to attending a previous Fashions in the Garden event, it must have been like going from Mozart to Metallica.

It was twelve years later, in 1975, that attempts were made to revive Fashions in the Garden. By then the main driver of the original concept, Helen Dunn-Rankin, and art director Edith Tadd Little, had died. Nevertheless, the show went ahead on March 5, with the combined efforts of sixteen Park Avenue shops and Garden Club members, under the direction of style coordinator Mary Lou Horras. Brick walkways around the amphitheater were re-laid, the azalea and daylily beds replanted, and the wrought iron grille work on the stage welded, sandblasted and repainted. Boutique items were displayed in the garden center, and Edyth Royal's School of Dance provided entertainment for the show. Tickets were $2.50. Subsequent shows happened in 1976, 1977 and 1979, when fashions by Jacobson's of Winter Park and music by the Manila Musikeros entertained a large crowd, with over 500 tickets being sold and $1,200 raised, not for the beautification of Mead Botanical Garden as in previous years, but for the Winter Park Garden Club Scholarship Fund. 1983 showed a profit of $1,656.26, and in 1984 only 300 tickets were sold but resulted in a profit of $1,348.12 for the scholarship fund. With declining attendances, the last show appears to have been in early April 1987.

5.6: *Helen Dunn-Rankin's essential contributions to the development of Mead Botanical Garden, and her pivotal role in establishing 'Fashions in the Garden' is recognized in a plaque at the base of the amphitheater stage.*

Fashion displays had changed considerably over the intervening years and were becoming increasingly commonplace. The big malls that opened in the 1970s, like the Orlando Fashion Mall, had regular displays each day of all the current fashions under one roof, so perhaps its time had come. But overall, Fashions in the Garden was a well-supported affair that generated significant income for the Garden and in later years for the Winter Park Garden Club. The charismatic Helen Dunn-Rankin was recognized as the originator and guiding spirit of the event. It was her vision and persistence, supported by loyal committee members, that got things done. In recognition of her services, the Winter Park City Commission presented her on November 4, 1959, with a small statue engraved "Mrs. FIG of 1959," with high praise for her services to Fashions in the Garden and the cultural growth of Winter Park. A plaque remembering her outstanding work for the Garden can still be seen at the base of the amphitheater stage.

CHAPTER 6

The Roller-Coaster Years

Once the City of Winter Park completed the ownership and maintenance of the entire Garden in 1955, the complaints bell started ringing in the city offices from local residents on the west side of the Garden, who were concerned about the poisonous coral snakes in their neighborhood. They believed the old claypit area was responsible for their numbers. Residents claimed that they had been asking the city to clean up the area for the past five years on account of it being a breeding ground for mosquitoes and snakes, and a fire hazard due to the tall grass. Eleven coral snakes had been killed in that area, eight of them recently, with the latest, a monster 36-inch one captured on the lawn of a home across the road from the Garden. It became so bad that fifteen residents drafted a letter to city commissioners demanding immediate action. The residents referred to this issue as a long-standing feud with city officials; however, city manager Clark Maxwell claimed this was the first time the problem had come up since he was appointed city manager 15 months earlier, and promised city crews would move into the area with disc harrows as soon as possible. But the angry residents couldn't wait and armed with sticks set about trying to flush out the venomous creatures.

When the crews got to work with a disk harrow, clearing under the weeds and grasses in the high-wooded area of the Garden, then the claypit region, they found not a single snake. The cleanup was extended to include all the swampy area to Pennsylvania Avenue to the east, with the final city conclusion, "We did not find any snakes, but workmen did scare off several rabbits and rats."

The high-wooded area of the Bartels tract was clear of snakes but still unused for garden purposes. This land had been given the description of undeveloped in the unsuccessful Orange County schools grab of 1953. In 1957, the city tree commission suggested the site as a tree nursery, where different types of young trees for use throughout the city could be grown and nurtured. The area chosen was to the right of the Garden Drive entrance, and that was cleared, beds prepared and planted, and a sprinkler system using deep well water installed. Eventually, there were around 175 young palm trees, 250 young flowering trees, and 750 live oaks that formed the primary stock in the nursery. Using raised beds, many young plants were added, annuals were started from seed, cuttings were grown with mist propagation, and young hedge plants brought to maturity. The tree and plant nursery supplied thousands of items to the many city park locations over the years and was finally moved to a new position in 1968.

As for the rest of the Garden, although there was still a 50¢ admission charge in 1957, maintenance issues were at an impasse with no funds budgeted for the work. McCaully, a member of the Parks Board, expressed the concern to city commissioners that there were "pressing needs at the garden to put it in first-class shape. Nothing has been done, and there is work that needs to be done now." As an example, he pointed to the fact that if the creek was not cleaned at this time, all the paths would be flooded when the rainy season started. The Winter Park Garden Club, now using the lodge as their meeting point and center, were equally concerned. To assist the city, they agreed to renovate the interior of the lodge and its immediate surroundings, in addition to selling potted plants to help with expenses. This followed a request from the Garden Club to use the lodge as a public information center and workshop and to house horticultural exhibits. The petition also included a request to waive the entrance fee to the Garden for members. The city agreed on the condition that the Garden Club members continued to assist in the planning and development of the Garden, and granted an expenditure of $100 on the lodge to repair plumbing and the roof.

6.1: Mac McConnell, was the longest-serving garden employee. From William Jess, the resident orchidist at the time when he joined in 1945, he learned the rudiments of orchid care and other aspects of horticulture, eventually becoming superintendent of the Garden. He retired in 1968 after 23 years of service.

With no part of the Garden now located in the City of Orlando, the main entrance was changed to Garden Drive, and two additional maintenance personnel, Ellis Foust Jr. and S. T. Gaines, were employed to work alongside Mac, the chief caretaker. Three months of extensive clearing began to show dividends: weeds were removed from the trails and the underbrush in the hope that this would uncover many of the valuable plants that had been hidden for years. Weeds and silt were removed from the two lily ponds and parts of Howell Creek, and two dams and spillways built in the creek to keep the lake levels up. The lodge was completely repainted and fitted with new windows and screens. Future plans were for the extension of Garden Drive, presently a sand-rutted street inside the park, to receive a firm clay surface when the current dry spell was over, and then lined with palms. A new path would be opened to Lake Lillian where the rookery was located, with benches for bird-watchers, and with Garden Club cooperation it was hoped to add as many new flowering shrubs and bushes as possible to assure year-round color. Other recommendations from the Parks Board in 1956 included the remarking of all trees and shrubs in the garden limits. With the aid of a local botanist, some 60 varieties had already been identified and new nameplates, protected by plastic, giving the common name, family name, botanical name, and plant origin would be produced and placed on each plant. In the longer term, the Board had the ambition of landscaping the gatehouse at the Garden Drive entrance, repairing all the greenhouses, redesigning and redecorating the Edwin O. Grover patio, landscaping the iron gate entrance on Garden Drive, and installing sprinkler systems.

At the beginning of 1957, the Florida Audubon Society approached the City with a plan to create a new headquarters building on the corner of Melrose and S. Pennsylvania Avenue, close to Lake Lillian. The plans called for a 26-foot by 39-foot building with access to Pennsylvania Avenue, erected on city property, with the society taking a long-term lease. The society's interest in this location stemmed from the adjacent area of the bay swamp which they suggested "be dedicated to the protection and development of the heron rookery which it now surrounds," and agreed to take responsibility for this task if the building were built. The dock on the edge of the lake was already supplied with chairs for nature lovers who wished to study bird life and the nesting activities of the herons and egrets, when the trees at the side of the lake were white with snowy egrets.

6.2: *The heron and egret rookery of Lake Lillian. In spring the trees on the far shore were white with nesting egrets.*

The Parks Board recommended the plans to the city commissioners, who tentatively gave their approval, subject to city engineers making sure that acceptable sewerage and utility accommodations could be assured for the building. The next record of the proposal is a flat rejection by the City in early October 1959, with the terse comment, "An appeal by Dario J. Icardi on behalf of the Audubon Society for a 15-year lease plus a 10-year option on city property in Mead Botanical Garden was rejected by the commission." No reasons were given.

The city commissioners were keen to develop the Garden site in other ways and were more favorably impressed by a request from the Florida Federation of Garden Clubs for a place for their new headquarters. They offered a three-acre site with a 600-foot frontage along Maitland Avenue and extending 220 feet into the Garden. The projected building would be of a size to accommodate meetings of the board and parking for 100 members. On January 6, 1960, Edwin Grover stood on a carpet of pine needles and watched dignitaries with a golden shovel turn the first earth for the new building. He must have felt a certain satisfaction after more than a quarter of a century of dedicated work creating and guarding the estate for the community and Central Florida. The $60,000 building opened quietly on September 1, 1960, with a formal dedication set for January the following year.

Just when everyone thought land grabs by Orange County officials were a thing of the past, after the 1953 debacle, the possibility of a new elementary school for Winter Park located on that part of the Mead property was reinvented and sprung on the unsuspecting guardians of the Garden in July 1959. This time the target land was not part of the Garden, but the Agnew and Nelson plots of around ten acres on the corner of S. Maitland and Melrose Avenues adjoining the north-west corner of the Garden. But the proposal also included the school taking the adjacent four acres of the claypits for playgrounds, and that this land would be traded with the city for the area currently occupied by the Park Avenue Elementary School. The argument centered on whether or not the claypits were a part of Mead Botanical Garden or not, since they were originally owned by Orange County.

A legal fight was once again in the offing. Arguments from both sides were heard, with the city parks and trees board unanimously voting that Mead Garden be kept intact for park purposes, pointing to the fact that 100 trees had already been planted there as part of creating an arboretum. Garden Club members and some citizens also protested about utilizing part of the Garden for the school, and Mrs. Agnew, the owner of one of the properties, said she would protect the Garden by not selling any part of her property for a school.

The proposal to take land from Mead Botanical Garden for the school was the second in five years; the first had been rejected five years earlier in a city-wide referendum. Now, with the question back again, there were some who believed that not only had the city neglected the gardens to the point where they were becoming valueless but that the city was trying, piece by piece, to get rid of them. In their opinion a much better school site would be the fourteen acres bordering the Garden to the east and north of the Dinky Line, currently owned by J. J. Banks, and a site enthusiastically endorsed for a new school by Edwin Grover.

Despite other city leaders praising the idea of the Banks' site, Winter Park Mayor Pflug favored using the four to five-acre claypit belonging to the Garden for ball diamonds and recreation, so the Orange County School Board persevered with the Nelson/Agnew location. Following an appraisal of the options, the school board dismissed the Banks property as being too far south of the optimum position and also near the boundary with the Audubon Park Elementary in Orlando. The debate dragged on, kept alive by some senior city officials, who wanted to use part of the Garden, and concerned citizens such as Edwin Grover and the Parks Board, who wanted the land left intact. The impasse was only broken when plans for a new apartment complex on the seven-acre Nelson tract appeared at the Winter Park Zoning Board meeting in January 1961, and was approved. Mead Garden Manor would consist of 200 apartments built by the Adler-Bilt Construction Co. selling for around $14,000 each, with construction starting soon.

Grover was getting pretty fed up with having to defend the Garden property from an increasingly cash-strapped and belligerent City of Winter Park. The school board fiasco was close to the final straw and he castigated the city in a newspaper article in July 1959, headlined "Dr. Grover Deplores Mead Destruction." In the article, he accused the city of "step by step destroying the Garden." He went on to suggest that the Garden had deliberately been allowed to run down and that he, on a professor's salary, had put more than $20,000 of his own money into the Garden. He concluded, "If they give it away, it can't be developed again the way it was planned."

The city's attitude to this once-beautiful garden was loutish and uncaring. Quietly and surreptitiously they had been using it for years as a rubbish dump, discarding city garbage and other debris in the claypit area. Waste was buried in the bottom of the claypit and when that was filled, deep holes were dug with bulldozers over the upper end of the Bartels tract which were then filled with garbage, creating such a stench that the people of Beverly Shores complained. When a brick-lined street in Winter Park was renewed with fresh bricks, the old ones were spread out and left in the Garden, as were chunks of concrete and curbing from demolition and waste materials from city building sites. Step by step the city really was destroying the Garden, and the Bartels tract reversion clauses were simply being ignored.

In March 1961, the city commission awarded a contract for a refuse transfer facility at the claypit to enable large items of waste to be transferred to big trailer trucks before being hauled away for disposal. Once again, Grover took to his pen, writing to the *Orlando Sentinel* how opposed he was to the refuse transfer facility, reminding the readers of the reversion clause, and that, in his view, Mead Garden had suffered sufficiently at the hands of the city officials. He concluded, "It seems inconceivable that the officials of beautiful Winter Park should use its largest and most beautiful city park for a refuse transfer facility."

Once the cat was out of the bag, and the real extent of the dumping revealed, complaints flooded in. When confronted, Mayor Edward Gurney cleverly turned the problem round so that it appeared to belong to the complainers, by asking them to propose another location that was as desirable from a city's point of view. On April 13, 1961, a charged-up group of Garden Acres residents appeared before the city commissioners, claiming that apart from the smell and desecration of the Garden, the trash facility would depreciate property values. Tempers ran high but most restrained their remarks, until a surprise punch was thrown by H. C. Buchanan, of 538 Camellia Avenue, who stated that the law firm of Thompson and Britt was looking into the matter of restraining the city from using the claypit. He reminded the commissioners of the reversion clause in the Mead Botanical Garden land grant from Mary Bartels and accused them of violating it, and finished by informing them that the lawyers were going to Jacksonville to visit Mrs. Bartels to see what could be done about it. There is no record of what happened next.

Garden Club members celebrated Grover's 91st birthday with a birthday cake and the dedication of a trail marked with a plaque in Mead Botanical Garden. He was interviewed on his 93rd birthday by the local newspaper that praised his community spirit in founding and developing the Garden with Jack Connery. When asked by the reporter what would be the best birthday gift to mark the day, he replied that he wished to see a renewed interest developed in Mead Botanical Garden by the public, and particularly by the city fathers who held the power to make the necessary improvements.

6.3: Edwin Grover cuts his cake on the occasion of his 91st birthday in June 1961, supported by Mrs. Helen Dunn-Rankin and Mrs. J. L. Saylors, president of the Winter Park Garden Club.

Although in his early 90s, Grover was still in a fighting mood when it came to matters of the Garden encroachment, believing that so long as the property was kept intact, there still was a chance that it could be built back to the garden of beauty that it once was. For the time being, he continued to do what he could whenever he could. A walk out of his house at 930 Camellia Avenue to the Garden could take ten minutes or so since he walked with a cane and was generally accompanied by his daughter, Frances, to make sure he didn't have a fall. Occasionally, Grover would take along his yardman if there were areas of high grass and weeds that needed cutting back, but he always had time to tend to what few orchids remained of a once outstanding collection, and talk to Mac, the caretaker.

Grover's love of orchids was shared by his friend Martin Andersen, who was first smitten by their beauty when he saw the Mead collection in bloom shortly before the Garden opened. Andersen helped bring countywide interest in the Garden, and in about 1960 he started promoting orchids and sending them as gifts from the *Orlando Sentinel* to new mothers and severe hospital cases. The newspaper cultivated the orchids at its ranch under 30,000 square feet of glass and sent orchids to patients throughout the year at most of the Central Florida hospitals, made possible by the large number of different orchids they grew, guaranteeing blooms year-round. The orchid business grew, and in 1963 the Sentinel-Star Retail Orchid Center at 1717 Edgewater Drive was officially opened. The nucleus of the new center was the nationally famous Vesey Orchid Collection which the *Sentinel-Star* purchased from David Vesey of Wakarusa, Indiana. Vesey's father founded the collection in 1895 and devoted his life to developing hybrids noted for their large size and heavy blooming characteristics. The expectation was that the center would be a tourist attraction where, without obligation, visitors could view from 100 to 150 flowering orchids.

6.4: *Edwin Grover, aged 93, examines the orchids with Bev Brown, City of Winter Park Parks Director.*

Grover was nearing the end of his road, but his dream of Rollins College getting involved actively with the Garden's development never left him. Over Christmas 1963, he tried one more time, appealing for financial help from one of the greatest philanthropists of Rollins College, Archibald Granville Bush. In the letter he sent, he described the Garden and its current difficulties, pulling no punches in calling it badly neglected and abused, with most of the orchids in the greenhouses having died through lack of care. He expressed the opinion that as long as it was under city control, it would never be developed because the city just didn't have the money and had ever-changing policies and personnel. Since Bush was known for his support of the sciences, Grover stressed the value there would be in operating the Garden as an integral part of the Rollins College science department used for training in botany, biology, and horticulture, and pleaded for support since "it breaks my heart not to see it further developed." Bush's reply has not been found, but presumably, it was a politely worded rejection.

Over his lifetime, Grover was no stranger to personal tragedy and on two separate occasions had tasted the bitterness of the loss of loved-ones. First in 1936, when his wife, Mertie, was struck by a car and killed outside their house in a tragic accident; the second, only four years later in 1940, when his only son, Graham, apparently committed suicide by falling under a train. Following these tragedies, the Garden had been his all-consuming passion in the second half of his life, and he had happy memories of the early years when it had thousands of visitors and dozens of flower shows. He never lost his love of green spaces, nor the belief that growing cities needed more, not less, park areas and that they should receive public funding to keep them beautiful. On November 8, 1965, with this particular belief unfulfilled, Dr. Edwin Osgood Grover, Professor Emeritus of Books at Rollins College and co-founder of the Mead Botanical Garden, died at the age of 95.

Six years earlier, another stalwart of the Garden had also passed away. Dr. Eugene Shippen, key philanthropist and perennial garden supporter, died aged 93 at his Casa Felice home at 1290 Park Avenue North, Winter Park, on January 11, 1959. Dr. and Mrs. Shippen were both keen supporters of the Garden, serving on the Board, donating money, and, in Mrs. Shippen's case, working tirelessly through the Winter Park Garden Club on FIG society business and many other activities. Eugene's donation of $1,000 in 1939, shortly before the opening date, was a lifeline at a time when the Garden was in real financial crisis.

6.5: With the death of key Garden supporter Dr. Eugene Shippen, in 1959, an area of the Garden down by one of the large lily ponds was dedicated as "The Shippen Retreat." Mrs. Shippen gave a carved lead Florentine figure of St. Francis and had it placed on a pedestal in the corner of the retreat in his memory.

In his memory, an area close to the Garden Club building and beside the waters of one of the lily ponds was created and dedicated to the memory of Eugene Shippen as "The Shippen Retreat." W. Taylor Briggs, City Parks Superintendent, and his crew cleared jungle-growth from the site and planted a sunken garden, with two garden seats beside a large seashell ornament. They built a wall from curbing taken from Morse Boulevard and formed steps with a curved garden path to make a peaceful and quiet spot. The planting was mainly ferns interspersed with blue and white flowers, the Shippens' favorite colors.

The dedication took place on Saturday, June 20, 1959, when more than forty people gathered in his memory, and heard Helen Dunn-Rankin dedicate the retreat in the name of the Shippens, who had given generously to the Garden through the years. Mrs. Shippen presented a carved lead Florentine figure of St. Francis to be placed on a pedestal in the corner of the retreat.

In 1959, Mayor Lynn Pflug called a luncheon meeting of seventy-five leading Winter Park business people to discuss how long-range improvements in Winter Park could be financed. Before the general discussion, the mayor made some observations, raised some questions, and floated some possible solutions. His address covered many of Winter Park locations including Mead Garden, which he referred to as a controversial project. He described the Garden as like an orphan left on the city's doorstep, and conceded that at least twenty gardens in the Southeast and Florida would make Mead Garden look like a weed patch. He alluded to the attitude of certain Mead Garden people towards outsiders and specific civic expansions in the area that he believed had alienated the affections of many people. But, he continued, despite personalities there was nothing so wrong that $50,000 or so couldn't cure. A committee of three key members concluded the meeting by suggesting a $2.5M issue to cover all locations and financed by a 1.5 mill tax rise might be acceptable to the people, so long as specific plans were drawn up noting the exact amounts to be expended on each improvement site. Nothing further was heard of this suggestion.

On January 1, 1960, access to the Garden became free, and the 50¢ admittance charge abolished. Unsurprisingly, this brought a flush of local visitors, which the city parks department claimed as some record and, as a result, increased its plan to install more picnic units. The entire fifty acres or so were put under the care of one general maintenance person and the long-servicing caretaker Mac McConnell, who lived in the gatehouse at the entrance from Garden Drive.

Back in 1945, David McConnell was flicking through the pages of the *Orlando Sentinel* when he came across an advertisement for someone to look after the flowers at Mead Botanical Garden. He was an ex-farmer, familiar with plants, and a keen and rapid learner, so he applied and got the job as caretaker, with accommodation at the Garden in a wooden shed by the entrance of Garden Drive. From William Jess, the resident orchidist at the time, he learned the rudiments of orchid care and other aspects of horticulture, eventually becoming superintendent of the Garden and in charge of the orchids in the greenhouses.

6.6: *Mac stands in front of his old wooden home in the Garden (left), now a dilapidated shed, ready to move into his new concrete gatehouse at the Garden Drive entrance (right). Dated 1957.*

The old wooden shed that was his home became so dilapidated, and judged a fire hazard, that in 1957 the Parks Board decided to replace it with a new concrete block gatehouse. An appeal went out for soft furnishings and furniture, and the city commission voted $200 for an electric range and refrigerator to complete the new accommodation.

Mac was a dedicated and loyal employee, whose seemingly simple exterior hid some very radical right-wing views. He had no time for liberals or labor unions and hated communists, calling them "red vipers"; he believed in the power of prayer and, as a teetotaler and vegetarian, keeping the body clean and pure. On one occasion, he was interviewed by the *Orlando Sentinel*, who described him as a "wiry old elf." Mac related his experiences to the reporter, remembering the bad times when Dr. Grover's age forced him to retire from the Garden management. At that time, just before the City took it over, "Too much inefficient help" according to Mac, was the cause of the near collapse of the Garden. When he first arrived, he explained, there were five regular employees, now there were just two.

Around 1956, when the Garden's future appeared shaky, Mac bought a property in Gotha, on Hemple Avenue, with a grove of twenty-five orange trees, and rented it out. After a six-day work week, he usually took Tuesdays off to visit his daughter in the area and check on his house and trees in Gotha. He reckoned that life in the orchid house would teach one thing if nothing else, and that would be patience. In 1968, after 23 years as caretaker, he retired from the Garden, leaving his deputy, William Ferrigno in charge of the orchids. He returned to live at his home on Hemple Avenue, dying there after a short illness, on December 18, 1987.

At the start of the 1960s, City of Winter Park Mayor Pflug, having failed to get more tax dollars for city-wide improvements, which included Mead Garden, called on people having an interest and spirit within the community to come forward in the form of volunteers and help revitalize the Garden. For their part, the city started work on the claypit area, shaping it to be a grassed play and picnic area, and planted two hundred hibiscus plants along the primary drive into the Garden, donated by the Mid-Florida Chapter of the American

Hibiscus Society. A collection of cacti, given by Mrs. Dorothy Gordon, joined the rest of the plants in the orchid house.

Newly appointed City Parks Director, Beverly Brown, was also keen to see some improvements in the Garden and in 1963 announced plans for revamping and expanding the orchid houses. He acknowledged that the original outstanding orchid collection had become neglected and was in disarray, but was hopeful that an exchange program with orchid fanciers in Winter Garden, Miami and together with the local orchid society would rejuvenate the collection. He admitted that years of neglect had affected not only the Garden, but the orchid houses too, and as a result, many plants had died, were stolen, or their name tags lost. With the help of several people, Brown's aim was to re-catalog the collection which an inventory of the three greenhouses revealed stood at 1,949 plants. Unfortunately, he added, the re-potting, tagging, spraying and fertilizing of the orchids would not be done.

Other improvements in the Garden were also underway. The intention was that the entrance from Garden Drive to the amphitheater would be brick-lined, and that a camellia garden on the north side of the main gate would be created. Brick-lining took some time, with old bricks available from the relining of existing city streets, and it wasn't until late March 1964 that the last brick was laid. The drive from the park entrance ran back about 1,200 feet, and it took an estimated 80,000 bricks to pave.

Brown instigated free public tours of the three orchid houses at 10–11 a.m. and 2–4 p.m. every day, with either Mac or William Ferrigno on hand to conduct visitors around. "Until now the orchid houses have been open all day, but due to recent inventory, cataloging and cleaning up these facilities, I feel that visitors will enjoy their visit more if someone can be on hand to answer questions and point out plants of special interest," he explained.

Brown was positive about the orchids, but his long-term view echoed that of the City of Winter Park – the garden transformation he was spearheading was from a botanical garden into a community park, which the city hoped to develop into a family retreat for weekend cookouts. To this end, several thousand dollars had been earmarked to be spent on landscaping the entrance drive, which at present was seen to be the most barren part of the park.

Even when there was an admission charge for the Garden, there were so many unofficial entrances for creative juveniles that occasional incidents of vandalism were inevitable. The Lodge, in use by the Garden Club as a garden center, was a target and in December 1957, intruders forced their way into the men's room at the north end of the building and then broke an 18 by 50-inch plate glass window to get into the main building. The padlock on the music room was pried open with a claw hammer, and two glass panes in the front door were smashed. Nothing, however, was reported missing. In May 1958, irresponsible juveniles chopped down several young oak trees in the Garden, something that goes on even today.

With the Garden being free, cases of vandalism increased, peaking in the late 1960s. In May 1966, one of the bridges was turned over and a large sundial in front of the Garden Club broken; in January 1967, vandals broke several globe lights in the amphitheater; and in October 1968, the Garden Club was the target again when the perpetrators broke down the front door, moved furniture around, scattered cigarette ashes and paper over the floors, emptied drawers and tore down a drapery partition.

Continued vandalism in the Garden and at the Lodge was reaching major proportions and costing city taxpayers an estimated $10,000 annually. In January 1969, a new plastic greenhouse that cost the city about $4,000 was ripped to shreds with a knife. Earlier, the plastic had to be replaced when someone walked across the top, punching holes in the plastic. Also, some concrete picnic tables were broken in half by throwing them on the ground and benches were tossed into the lake. The fate of the small lead statue of St. Francis on a pedestal in the Shippen Retreat was a more mysterious one. Having graced a quiet corner of this shady nook for more than eight years, it disappeared one day in 1968, only to be returned ten months later in early January, 1969, in perfect condition and even a little cleaner than it had been. Having got the statue back, it was secured to the pedestal by pouring cement around the base. The rejoicing over the statue's return was short-lived, however, and a week or so later it had been torn from the cement base and disappeared again, this time never to return. All that remained was the brick base, and a metal pole which ran up the middle of the statue to hold it in place.

The amphitheater area, built and beautified by the efforts of Garden Club members under the leadership of Helen Dunn-Rankin and Elizabeth Tadd Little, and the Fashions in the Garden Society, was getting increasing use too. Music filled the skies in February 1960, as the Winter Park Civic Orchestra performed to more than 500 people in the first of three planned community concerts. Under brilliant, though chilly, sunshine, Conductor Arthur Bell led the orchestra through a medley of the old and loved; among the favorites were the Mexican Overture by Merle J. Isaac, and the Slavonic March by Tchaikovsky. A number of other symphonic and community events were held in the amphitheater over the years, many sponsored by the Winter Park Chamber of Commerce. In 1967, it was the turn of the Florida Symphony Orchestra to play under the stars, under the baton of Yuri Krasnopolsky, and give an evening concert of light classics on March 29 and April 16 that attracted over 1,000 people. The crowd filled the amphitheater benches, but many had brought their own chairs or sat on the ground.

Big, grand, and expensive master plans were a feature of the Garden throughout its lifetime. The 1967 master plan was a classic example. Mayor Daniel M. Hunter, who admitted that the city-owned botanical showcase had slipped during past years, touted the plan as one that would make it "the finest such garden spot in the nation," a clear echo of the 30-year old declaration when the Garden was being built in 1937, namely "to be Florida's finest garden spot."

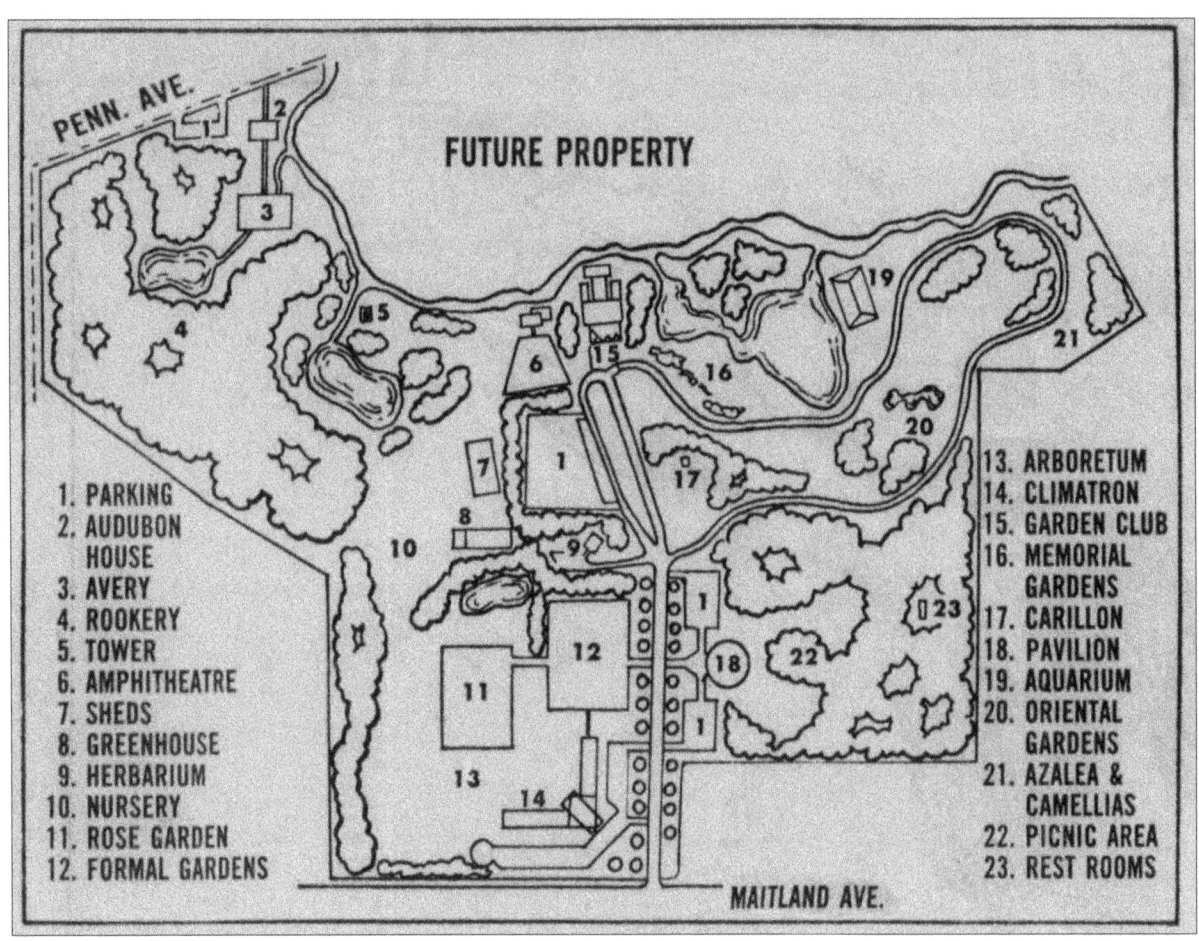

6.7: The 1967 grand master plan was finally unveiled to the public in 1971.

Massively ambitious, the plan called for a development over a five to six-year period of a large L-shaped greenhouse, or climatron (budget $55,000), to the left of the present entrance, to include a display area for special plant and flower shows; a bird sanctuary and aviary (budget $50,000); carillon and aquarium (budget $90,000); an Audubon house and observation tower; several new parking areas; a large picnic pavilion and several new roads, along with an enlargement of the Garden Center. It also assumed that through compulsory purchase the 20-acres or so of prime plant-growing land to the east of Howell Creek would be secured as part of the expanded garden. The five-year budget programs totaled $435,250, out of which land acquisition for the property east of Howell Creek was $225,000.

The consultants produced a plan that they believed "would give Mead Botanical Garden the proper and long-awaited respect it deserved and an asset of untold worth to the City, capable of gaining wide recognition as a Botanical Garden and Arboretum."

CHAPTER 7

Neglect and Destruction

It came as no surprise to most people when the $500,000 Mead Botanical Garden five-year master plan, capable of giving the Garden "wide recognition as a Botanical Garden and Arboretum," fizzled out and was quietly shelved. With no appetite for borrowing nor tax increases to find the money, and an increasing amount of public apathy, it was a non-starter. Maybe a sum of one-tenth of that for specific quantifiable improvements might have stood a chance, but aviaries and aquariums didn't make sense to the majority of Winter Parkers, and a vocal minority of the community anyway wanted it kept "just as Nature intended." Conceptual drawings of the planned carillon, aviary and aquarium were relegated to the back corners of the filing cabinets in the Parks Department offices, hidden from view.

In addition to the plan dying, the long-standing desire to make the land between South Pennsylvania Avenue and the eastern shore of Howell Creek part of the Garden was dashed, following the announcement in 1967 by the Seaboard Coast Railroad of the imminent closure of train services over the Dinky Line, with the physical removal of the tracks. Rights-of-way were granted to owners of the land traversed by the line, and this immediately created a sizeable prime housing development target to the east of Howell Creek, owned by J. J. Banks. He wasted no time in applying to build a series of high-rise apartments, townhouses and single-family dwellings on the land. Rejection of this proposal led to an alternative plan that was approved; to build luxurious family homes on large lots featuring Spanish style architecture and landscaped gardens, some enjoying views across to the Garden. The grand opening took place on November 15, 1970, with the offering of 63 building sites, nine of them lakefront, ranging in price from $17,500 to $45,000. In keeping with the Spanish theme, streets were given appropriate names such as Barcelona Way, Santiago Drive, and Salvador Square. The land that had been labeled "future development" for the last thirty years on maps of Mead Botanical Garden became the Sevilla subdivision.

Earlier, in 1966, before the decision to remove the tracks had been made, a proposal had been submitted by the Orange-Seminole Joint Planning Commission to use the Dinky Line as a kind of recreation linear connector between many of Orlando and Winter Park's attractions. Starting at Orlando's Loch Haven, visitors could link via the railroad with Mead Botanical Garden, Rollins College, Genius Drive, and the Winter Park chain of lakes. The proposal called for an extension of Mead Garden, but one of the most significant expansions suggested was the creation of Genius Park in the Genius-McKean Estate, an area of about 270 acres of land not presently owned by the city. The vision also recommended putting a pedestrian trail alongside the Dinky Line and operating a small-scale sight-seeing train pulled by an old-fashioned engine, with open-sided cars, through the different attractions. Winter Park Mayor-elect Dan Hunter thought the idea "fascinating" with "tremendous merit," and mentioned that Ward Park might also be included in the recreation facility. He proposed that a committee of business people evaluate the concept as to its feasibility and then report back.

The timing of the proposal coincided with the election of a new Winter Park city commissioner for the vacant site created by Dan Hunter's appointment as mayor. Three commission candidates were asked at a meeting what they thought of the Dinky Line Park proposal, and all three poured scorn on the concept. They suggested that the use of the Dinky Line in a giant park wasn't for Winter Park, with one going so far as saying that it wasn't an intelligent idea and would never be a reality. But attacks on the proposal were termed "premature" by the Winter Park Chamber of Commerce, stating that it could be "an embryo of a merchandising idea" which could be used to display the community, and deserved serious consideration. However, several other members of the committee labeled the plan "utterly ridiculous," with one member believing that most residents thought the plan was "hardly feasible." No more was heard of the proposal.

The Winter Park Garden Club had been operating out of the Reception Lodge as a meeting center since the mid-50s, and began formally leasing it from the city in 1961. Having secured the lease, they commenced a funding drive to replace the old wooden structure with a 45 by 134-foot building of natural wood and glass, having an auditorium to seat 600 people. Donations had started coming in towards the target $40,000 needed, with $5,000 already in the bank. The lease contained a reversion clause stating that the city had the right to purchase the building if the Garden was misused, which had the members of the club chuckling. Having hand-spaded and dug the Garden out from the jungle over the years to make it what it was today, they naturally asked why on earth would they be the ones to misuse it?

Over the years, funding slowed to a crawl and the initial expectation of an opening in the fall of 1961 faded. In 1969, a new funding campaign was launched for a modified building designed by Clifford Wright, a Winter Park architect, having an auditorium with a seating capacity of 350, a gardening library, a kitchen, storage rooms, and an extra meeting room. Some of the original building's pecky cypress would be salvaged and used in the new building's cladding in a herringbone pattern paneling. The exterior would be of stone and block masonry, and glass areas would be kept to a minimum to reduce vandalism and make it easier to cover when slideshows or movies were shown.

A few years later, in May 1971, the fund stood at $43,000, but by then the building costs had risen to $50,000, and it would be November 1972 before the Allen Trovillion Construction Company broke ground and started building.

Dedication of the new structure and the official opening, assisted by Kenneth Murrah, Chairman of the Winter Park Chamber of Commerce, took place on September 23, 1973, with the theme "A Dream Come True down our Golden Pathways to Gardening Activities."

7.1: The new Winter Park Garden Club building, replacing the original wooden reception lodge, officially opened in September 1973.

In the early 1970s, valiant attempts were made by the Winter Park Parks Department and concerned volunteers and citizens to preserve at least part of the floriferous nature of the Garden. By then the camellia collection had been moved to the western edge of the Garden, but there were still daylily plantings and some orchids in the greenhouses, looked after by William Ferrigno and caretaker Lee Simpson.

Jay Blanchard, Parks and Recreation Director, had joined the department in 1965 and had a useful horticultural background. He was of the opinion that Winter Park needed more parks, not less, but believed that ideally a number of small, individual, well-developed parks connected by walkways was preferable to one big park for everyone. With this arrangement, people could walk from one park to another, stop and rest on one of the benches and enjoy the flowers and plants, before moving on to the next park. In 1972, he requested an additional five employees for the parks department, arguing that many existing parklands in the city had not been developed to their potential and this was the first time he had asked for employees in eight years. The request was turned down. Unsurprisingly, Blanchard resigned in early 1973 to take the job of Parks Director for Orange County, and this proved part of a cascade of resignations of people from senior positions within the city administration.

More chaos in the administration followed the decision made by the city in August 1973 to place parks under public works and to make recreation a separate department for the next fiscal year. This division left James Buck Jr., Jay Blanchard's assistant for three years and his successor as director of parks and forestry, in a quandary. Buck had joined the department in April 1971 and was in charge of about 60,000 plants in the city nursery, the horticulture and spraying of city plant material, the acquisition of parkland, and the design of park facilities. He was a professionally trained parks and recreation administrator and could not understand the logic of separating the two, so he also resigned.

In his letter of resignation, he registered his disappointment with the split and said that he had taken the job initially because he believed Winter Park had one of the most progressive parks and recreation departments in the southeast, concerned with new park facilities, better-designed playgrounds, and the acquisition of more land to serve the residents now and in the future. The most progressive cities throughout the nation put parks and recreation together as one department, but now, he believed, as part of public works, maintenance would be the sole focus. He finished with, "I feel that it would be unethical both personally and professionally to administer a program in complete opposition to that of a professionally trained parks and recreation administrator." Buck left and became director of Seminole County's parks and recreation department, a job with more authority and a better salary.

7.2: The were still flowers in the Garden and orchids in the greenhouses in the early 1970s but things were about to change dramatically. Left: Jay Blanchard, on the left, talks to James Buck, both of the City of Winter Park, surrounded by flowering annuals outside one of the many greenhouses in 1971. Right: Betty Mobley tends the orchids in the last known photograph of the collection, April 1974.

Three more resignations of key city administrators followed, but all these actions were dismissed by the acting city manager, Bob Proctor, as usual for the time of the year when employees became uncertain of the coming year concerning salaries and department allocations, and did not indicate any morale problem. But it was clear to everyone else that there was a decline in morale, aggravated in part by the recent departmental reorganization.

The disastrous split between parks and recreation didn't last long. It was not popular and the collective view was that it encouraged a breakdown in communication between the departments and the city commission. The call to re-consolidate parks and recreation was loud and clear, but it could not happen until the start of the next fiscal year, in October 1975. In the following year, the city was forced to eliminate three and a half employee positions in an attempt to balance the budget; three from the Streets Division, but one full-time worker in the Parks Division was reduced to part-time work. In 1977, further cutbacks resulted in there being only one caretaker position for the entire 50 acres of the Garden, adding to the already mounting pressure on maintenance.

At least one of the greenhouses was still open in 1973 as a picture in the *Orlando Sentinel* proved, but judging by the photograph there were few orchids and the greenhouse looked run-down. William Ferrigno, who had been in charge of the orchids, retired in 1972, and orchid care in 1974 shifted to Betty Mobley, a city employee from the Lake Island Recreation Center on Harper Street in Winter Park, previously in charge of lawn bowling and shuffleboard courts for seniors. The last mention and photograph that has been found appear in the *Orlando Sentinel* of April 17, 1974, showing Mobley tending what was left of the Mead orchids in one of the original greenhouses.

The halfhearted attempts made in the early 1970s by the City of Winter Park to retain at least a small component of botanical nature in the Garden came to a halt, and from then on, the slow decline of the Garden accelerated. The decision appeared to be that there would be no development of Mead Garden, only maintenance at minimum cost. The city had stopped spraying, fertilizing and re-potting the orchids back in 1963, under Parks Director at the time Beverly Brown, essentially signing their death certificates. Now, in 1976, the orchids that were still alive and hadn't been stolen, about 300 of them from the original several thousand that provided an outstanding display in the 1940s, were moved for safekeeping to a remote location where the public could not view them. For some, it was in the hope that someday the city would build a showcase in the park for them.

The city tapped into free labor available from the Youth Conservation Corps (YCC), paid for by Orange County through funding from the US Department of Interior, Department of Agriculture-Forest Service and the Florida State government, to help clean-up the Garden. This scheme provided ten boys and ten girls aged from 15 to 18 from selected high schools in the county with employment for six weeks, working in the Garden four days a week and earning money and academic credits. Working alongside crews from the city, the focus initially of work in 1976 was on rebuilding some of the bridges and re-laying the trails, but this soon developed into the arbitrary moving and removal of plants. At this point, it appeared that the daylilies were dug up and the formal rose garden containing 650 roses close to the main greenhouse plowed up and the plants either destroyed, stolen or moved to other parts of the city. The Parks and Recreation records of the City of Winter Park relating to this entire period are mysteriously missing but there is some Winter Park Garden Club evidence that has survived that indicates what was happening.

7.3: In this scrapbook photograph, with its original caption, the Winter Park Garden Club appears to have captured the start of the flower and plant destruction of the mid-1970s.

The year 1977 brought a cutting rebuke from Sherry Andrews, writer for the *Orlando Sentinel*, and a broadside of complaints from residents over the state of the Garden. In May, the newspaper described the Garden as having little resemblance to the one described in the publicity brochure of the 1940s. It pointed out that the greenhouses had been abandoned and the orchids all gone, and that those greenhouses still standing were filled with rubble, empty pop bottles and cigarette packages, with their windows smashed by vandals. The daylily collection and the rookery were gone, and in answer to the question of what had happened to them, Assistant Parks and Recreation Director John Holland, who was in charge of Mead Garden, said he could not remember when either existed. Holland, whose office was at the entrance to the park, said he bore the brunt of visitor's disappointment. "People come into my office all the time irate about the condition of the park, but there's really nothing I can do about it," he said. "With one person responsible for 55 acres, there's no way to maintain the park in the way it should be. About all he can do is a little watering and clearing out the worst of the debris." Holland also mentioned the problem with theft, saying, "People come through here and just take whatever they want. They know there's only one person here and if he's at the other end of the park, they're safe."

Visitors from other cities were not the only ones concerned about the condition of the Garden, and many long-term residents of Winter Park also complained about its deteriorating state. Charles Sheppard, a supervisor of the Orange County school system's Environmental Education Center, called it rundown and the problem, "a matter of (lack of) civic pride." Members of the Winter Park Garden Club were equally concerned. "It's a beautiful, beautiful park if the city would maintain it, but it isn't up to par," said Harriet Fives, former corresponding secretary for the club.

In the midst of this maelstrom, Winter Park Mayor James Driver appeared to be in smug denial. Despite overwhelming opinions and evidence to the contrary, he clung to the belief that everything was fine, and would not accept that the park was neglected, claiming that, "It has just been left in its natural state, and most people do not want a natural park to have a well-manicured look." Where he got his "most people" from is anyone's guess, but it was the city's defense. After the usual excuse of personnel shortage due to lack of funds, and in a masterly stroke akin to marking his own homework, he declared that "I think overall the parks are well taken care of."

7.4: *In these photographs, taken from 2010, the neglect and deterioration of parts of the Garden, first noted in the late 1970s, is apparent.*

The Winter Park Chamber of Commerce Beautification Committee, concerned about the complaints, made an investigation of the park and filed a report to city officials, concluding that the park had obviously been neglected and had deteriorated. They met the "natural state" comment of the mayor by pointing to trash littering the gardens, abandoned greenhouses, and greenery declining to jungle status, as evidence of neglect.

An increased workforce of forty Youth Conservation Corps youths continued the clean-up over the summer of 1977 under the direction of crews from the city. The abandoned greenhouses were systematically demolished. The students built-up the nature trails, repaired bridges, planted azalea bushes, cleaned streams and put in new benches. Overgrown weed-filled areas were cleared, and, in all probability, what was left of the caladium and amaryllis collections were dug up. The rationale of the city seemed to be that it wanted a low-maintenance community park, that could be maintained by one full-time city employee, and flower beds just got in the way of achieving this. When the work was complete, the Garden was clean, tidy and landscaped, but barren of beauty. What became of all the plants, with the exception of the camellias and the few remaining orchids, is not known and there is no mention in the files of the City of Winter Park as to their fate.

Directors of the Youth Conservation Corps program offered to do additional work in 1978, if the city commission approved appropriate projects and if federal grant money was available. The projects would be completely funded by the Youth Corps, with the city providing some in-kind services. Grants for two projects were approved – a 300-foot raised boardwalk to protect the park's ground cover and wildlife from damage caused by visitors making new trails through the underbrush; and a plastic greenhouse to replace the five demolished structures and display the orchids to the public in a well-protected and visible area.

7.5: Top: In 1976, the orchids were moved offsite out of public view, and the magnificent Porter greenhouse, together with the other four, were systematically taken down. Middle: In 1978, Youth Conservation Corp youths erected a new greenhouse by the Garden Drive entrance. Bottom: By 1982, this structure was badly in need of maintenance and contained just a few tropical plants and one or two common shop-bought orchids.

Both projects were completed over the summer of 1978. The boardwalk ran into the wetlands from a point near the South Pennsylvania Avenue trail entrance toward Lake Lillian, and the plan was to use it for bird watching and environmental studies. Plans for the boardwalk area included spreading of wood chips on the lot off S. Pennsylvania for parking, and extending the boardwalk further into the park. A second YCC group built a 40-foot square plastic-covered greenhouse located just inside the main gate, with the intention of school children using it as an environmental study training center, once it was populated with plants that the city intended to provide.

Nature took over once the forty or so students left at the end of the summer, and the three years of summer work by the students started to look untidy again. By 1980, the call went out once more for volunteer labor to help, and a group of local scouts became involved with an Eagle Scout service project, involving painting the park benches and amphitheater stage and generally tidying up. The city was reluctant to spend any money on the Garden if it could help it, so once again, in 1982, it sought help in a campaign to revive the garden. By then the newly erected greenhouse was looking empty and sad, with leaves piling up on the roof, and containing a few shop-bought orchids that did not impress anybody. What happened next would ensure that what was left of the Mead orchids would never be seen again.

The Winter Park Chamber of Commerce's beautification committee continued to criticize the city over the condition of the Garden, asserting that this once nationally-known botanical garden was facing a severe identity crisis, and the few visitors it received were appalled by its disrepair. Berta Hall, of the committee, said that people used to pay an admission fee to come and see the flowers and plants. Today they drove in and became upset by what they saw – rotten planks and rusty leaning lampposts in the amphitheater area; outside walls of the bathroom facilities covered with graffiti and spray paint; an unsteady looking gazebo in the central park area; brick paths barely visible above the grass and brush that had grown over them; and weed-covered ponds in need of reclaiming and cleaning of sediment. The committee decided that something ought to be done and initiated a project to restore the Garden and renovate its greenhouse. To ensure the success of the project, the support of the entire community would be needed; citizens were encouraged to take an active part in rejuvenating the gardens, and those that did would become The Friends of Mead Garden.

They hit on a cunning plan to raise money for the upkeep of the Garden. With the best intentions, but in an action comparable to selling off what remained of the crown jewels, the Chamber of Commerce decided to auction off the remnants of the rare Mead orchid collection, which was hidden away in the maintenance yard of the parks department. They used the orchid expertise of personnel from the Central Florida Orchid Society to bring the plants back to a presentable state, and the auction went ahead in the greenhouse on May 22, 1982. Around 300 orchids were sold and $900 raised, an average price of $3 per orchid and a far cry from the $500 Theodore Mead was once offered for just one of his hybrids back in the early 1930s.

The Chamber of Commerce initiative had spurred the city crews into action, and they joined in with the activities of the volunteers, much to the appreciation of the beautification committee, who welcomed the change. "We have a lot more cooperation and interest from the city. They've come around quite a bit," said Sonia Wall, who added that until last year, the city's attitude was "negative" about committing funds to park upkeep. City workers lined the banks of Howell Creek with sandbags to keep them from eroding during the rains, removed fallen trees from walkways, restored brick paths and mowed overgrown open areas, but even with the city's help, volunteer efforts remained a keystone of the Garden's maintenance.

The beautification committee formed a separate Friends of Mead Garden fund, selling T-shirts at the Farmers' Market and other locations, and raised more than $2,000. In 1983, they were recipients of a $1,000 award from the Walt Disney World Community Service Program. They built support through membership, believing that an organized group would have more clout with the city than a series of individuals, and had a vision of bringing back a much more extensive collection of orchids into the greenhouse by the main entrance on Garden Drive. The group also wanted to improve the accessibility of the Garden to Winter Park residents and suggested something that should have been done years earlier. They proposed to the city that funds be included in the 1984 budget to construct a curb cut on Melrose Avenue so that a small, mulched parking lot could be built at the corner of Melrose and South Pennsylvania Avenues, making a convenient back entrance of the Garden, and one most accessible to Winter Park residents and Rollins College students. A parking area, shelters, benches, and landscaping would give people a place to enter the Garden using the boardwalk constructed with YCC labor that ran from there to Howell Creek. Berms, or small raised areas, along Pennsylvania Avenue, would be planted to soften the appearance of the parking area from the street. The city was unwilling to spend anything on Mead Garden, so nothing became of this idea.

The decade of the 1980s had many similarities to a time loop as portrayed in the movie *Groundhog Day*. Every year, volunteers made immense efforts to improve the Garden and tidy up, then the city responded with an initiative of its own, and then nothing happened. The next year this pattern was repeated, and again and again over the following years.

1985 was thus a rerun of every other year. A facelift by the city to the amphitheater by painting the benches green and hanging baskets of flowers from the lampposts, and the cleaning and renovating of the dressing rooms, constituted an improvement so that groups using the amphitheater could be charged $75 for three hours. The volunteer effort that year was spearheaded by the Orange County environmental educator, Alphonse Reithinger, who organized for school groups to mark a self-guided nature trail, restore one of the boardwalks, and create a booklet to inform visitors about plants and birds. In addition, the orchid house near Denning Drive was chosen for renovation by a horticulture class from Winter Park High School. Plans called for creating pathways and a waterfall using bricks and rocks the city had accumulated from other projects.

Work started in January 1986, with the involvement of members of the Winter Park High School's Future Farmers of America Club. They were looking for a community project to enter a national competition called "Building Our American Community," and asked the parks and recreation department's permission to renovate the 1978 YCC-built greenhouse. The city provided the materials and the students the labor; generally around 15 of the 60 club members worked every Saturday on the project, and the plans called for a sunken fountain, rock waterfall, circular brick walkway, and concrete footbridge. Bill Carrico, Parks and Recreation Director, was pleased that it cost the city much less to have the students do the renovation than to hire a company. "We've probably spent about $150 so far. Our biggest expense will be about $200 on a pump for the fountain and the electrical hookup for it, the heaters and the light. For what we're getting, if we contracted it out, we would be looking at about $8,000," he said. To finish the project off, the Chamber of Commerce donated $833 to buy a heating and fan cooling unit for the orchid house, and the city promised a changing display of blooming orchids would soon populate the greenhouse, but no record of this ever happening has been found.

It was the same old story in 1987, with volunteers trying to rescue the Garden from neglect and the unchecked assault of Nature. The volunteers, a group of Rollins students and Winter Park residents, spent six hours one Saturday in April on the first stage of a long-term effort to restore the gardens. Upon arrival, they found the brick walkways and concrete steps covered with leaves and dirt, the grass around the lakes spread with litter, and in the lakes and streams all kinds of debris, such as chunks of concrete, metal, old furniture, cans, and paper. With Saturday's cleanup behind them, the next project was to repair the boardwalks that ran through the park's swamps, since many sections had fallen away or were rotten. The volunteers were worried, however, that the chemical drums, used to support the boardwalk, might have contained toxic chemicals, traces of which could get into the streams that run through the wetlands. A second cleanup, by Rollins students and others, took place in November with the focus on the wetlands, picking up cans and garbage. In response, the city added new trash containers throughout the Garden and some benches near the lake. Litter scattered across the unfenced area on the south side adjacent to Nottingham Street was also tidied up by the volunteers. This two-acre land area, originally deeded by Walter Rose and then part of the City of Orlando until annexed in 1955, became a target for development in 1988 when local property developer Allen Trovillion applied to build houses there.

Trovillion asked the city to consider selling that piece of the Mead property so he could develop a small cluster-home community. Although the deed restrictions called for that land never to be fenced and always maintained as part of a botanical garden, Trovillion believed he could persuade Stella Rose, widow of Walter Rose, to remove the restrictions and by so doing, he said, the city would be ridding itself of a liability. The proposal was quickly rejected but pointed to the problem the city was having with the deed restrictions. In a strictly legal sense, the city had for the last twenty years or so been in breach of the limitations by failing to maintain the Garden properly and allowing it to decline to the point where it ceased to be a botanical garden. Whether this was ignorance of the wording of the deeds or just everyday convenience, we shall never know.

More appropriate building development in the Garden was suggested in 1991, when the Winter Park Historical Society had the idea of building a Florida-style cracker home as a museum at the south-west corner of the Garden near its entrance on Denning Drive. The association was facing a choice between a brand-new building or relocating a donated historical home designed by James Gamble Rogers and was split down the middle on which way to go. The new building group hoped to fund an initial building of about 2,500 square feet on the Mead Garden property that would cost about $150,000 and could be expanded when more funds became available. City staff recommended the Bartels tract as a possible building location since it had less stringent deed restrictions than other parts of the Garden, but warned that it did not explicitly allow for a museum. They then took another look at the deed restrictions and decided that they did not need the written consent of all of Bartels heirs. The society continued to dither but started serious fund-raising in April 1993 with the sale of map reproductions of Winter Park in 1908, taken from a watercolor image by Ray Trovillion, depicting the then 95 homes in the city. The copies were priced at $50, and 500 were available to buy. The sale coincided with the city offering the society a new home in part of the soon-to-be-renovated Farmer's Market at Lyman and New York Avenues, that was accepted by the society, and provided a place to store and showcase Winter Park's historical past.

Since the Garden opened in 1940 there had been sporadic sightings of alligators at Lake Lillian. Like the fish that got away, the narratives tended to grow with the telling in terms of how big the gators were. Sightings in the 1940s concluded that the lake was home to a family headed by an eighteen-foot alligator, as measured by military binoculars. A 1955 sighting talked of a fourteen-foot one, but it wasn't until 1964 that the beast became a menace to visitors. Richard Hunt, of Maitland Avenue, was taking a quiet siesta on a bench at the side of the lake when a seven-foot alligator came charging out of the water and chased him out of the area. This incident joined others on a lengthy police report, including a recent one when probably the same gator chased a boy and his dog. The critter was finally captured by Captain Cullifer and Officer Smith of the Winter Park Police Department, after a battle of an hour and a half. Owen Godwin, owner and director of Gatorland, arrived with an assistant and they put a noose on the mouth of the gator and took it to the gator sanctuary in the back of a station wagon. He estimated the age of the gator at twelve years, and a search of the area revealed that the bench was between the lake and the nest of the female alligator where numerous young were found.

It was more than twenty-five years before another large alligator in the lake started causing problems, when there was a report of one attacking a park visitor's dog on July 12, 1992. The gator had been spotted intermittently over the years but hadn't given cause for concern since it lived in a part of the wetland off-limits to visitors. Terry Perlier, a trapper licensed by the Florida Game and Fresh Water Fish Commission, was called in to capture the animal, assisted by Karl Lotspeich.

The 180-pound creature put up quite a struggle while local television crews filmed Lotspeich up to his waist in water, wrestling with the gator. Finally, they got a rope around the animal, secured the jaws with duct tape, and managed to drag the creature by the front legs out of the water and along a sandy trail to the waiting truck. According to Perlier, the gator was around fifteen years old and eight feet long, and in an area where it could have caused a lot of problems for visitors. Lotspeich lobbied state game officials to spare the animal, which they released into Lake Monroe in Sanford. In exchange for letting her go, Parks and Recreation Director Bill Carrico agreed to raise the $500 the trappers would have received for the animal's meat and hide.

7.6: *Alligators in Lake Lillian were sometimes troublesome and had to be captured. In August 1964, a six-footer was apprehended (left); in 1992, an eight-footer (right).*

In January 1988, the Winter Park Kiwanis Club had planned to rebuild the boardwalk into the wetlands, which was unusable in many areas because of missing or deteriorated sections. These plans were put on hold when it emerged that a city task force had been visiting the Garden, and proposed the creation of yet another master plan to revitalize the now neglected botanical garden.

It was a grand scheme in timescale and cost – twenty years and a few million dollars was the original estimate, and it appeared that nothing had been learned from the previous failure of the 1967 master plan, many years earlier. History should have told them that relying on donations to raise the $1.5 million to fund the changes and on a large task force of volunteers as the labor force to maintain and improve the Garden, were both forlorn hopes.

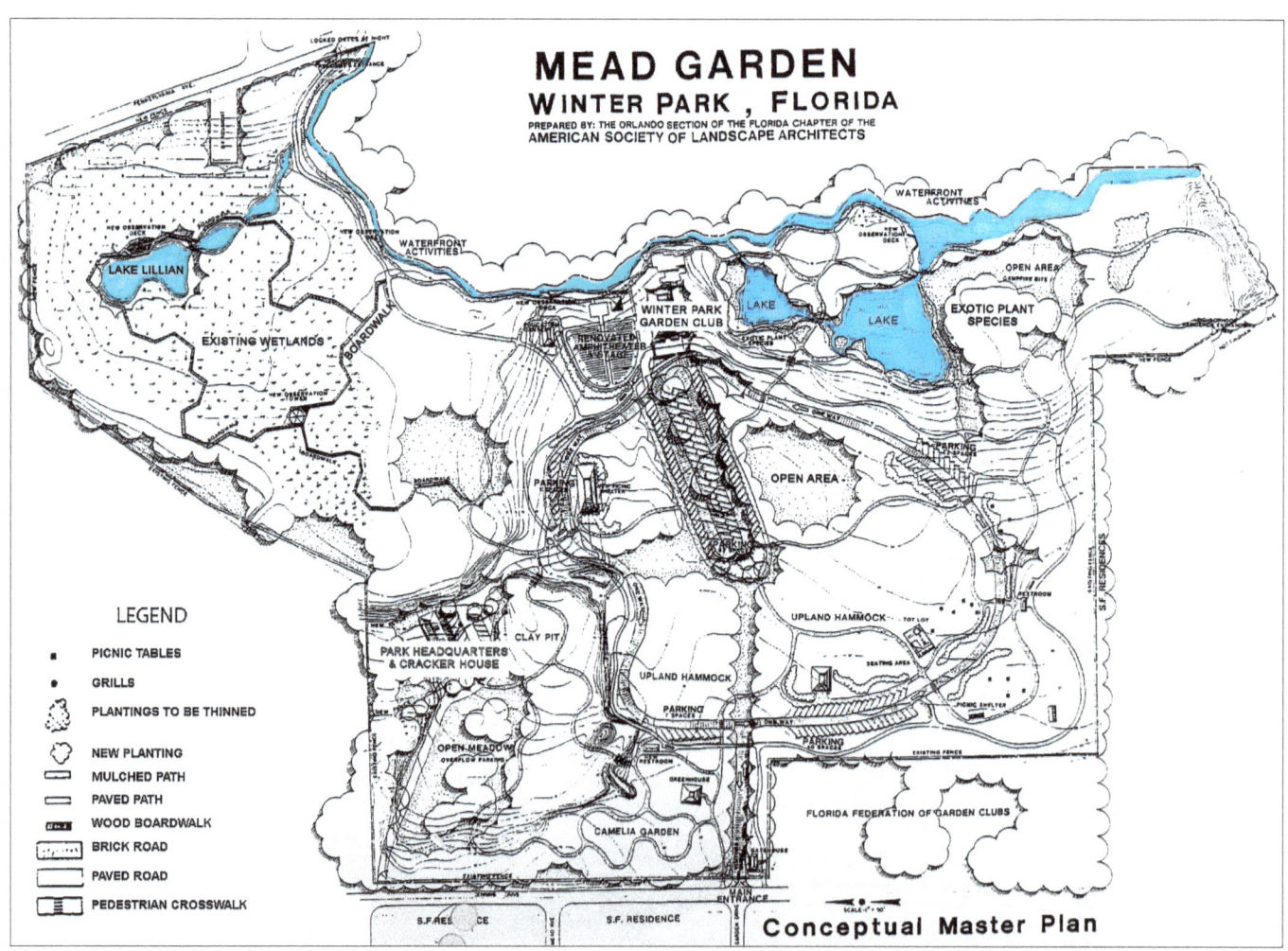

7.7: *The conceptual master plan of 1989. This time aquariums and aviaries were missing.*

But the plan, created by members of the American Society of Landscape Architects, had some first-class features – a circular boardwalk that would loop through the wetlands with identification markers turning it into a self-guided walking tour; elsewhere, areas designated for exotic plant species and well-marked trails and educational markers identifying trees, flowers and shrubs; a restriction on where cars and other vehicles could drive and park; and the installation of gates and entrance closures at night to prevent much of the current illicit activity by drug dealers, vandals and vagrants. An observation tower, picnic area, bike paths, and a jogging trail were other proposed features, and the plan was enthusiastically endorsed by the city commission and by a citizens' input task force.

The city once again pointed out that the money for these large-scale changes would have to come almost entirely from donations because Winter Park's recreation budget was strained just maintaining its parks, and there was no future intention, once the changes were completed, of increasing the workforce at the Garden beyond the two people already assigned. On January 14, 1990, to mark the 50th anniversary of the Garden's opening, the volunteer group, and the architects of the ambitious plan to restore the Garden kicked off a fund-raising drive to raise $1.5 million for the proposed improvements. The Mead Garden Preservation Association, created to spearhead the drive, presented its vision of the Garden's future in a slideshow and conducted tours of the property in the afternoon, and waited for a rich Winter Park philanthropist to come forward.

CHAPTER 8

A Flickering Renaissance

The 1989 master plan was made public in several Orlando periodicals and the boardwalk feature attracted much attention. Some preparatory work had been done in June of that year by a group of twenty-three Boy Scouts and Explorers from a program called Operation Comeback. They, along with a few adult volunteers, had hacked through dense brush while wading in waist-deep muck to clear a blocked stream that would help drain the flooded wetlands, ready for the planned future boardwalk and nature trail that would loop through the area.

In March 1992, the Winter Park Lions Club presented the Mead Garden Preservation Association with a check for $1,000 so that part of the boardwalk could be constructed and designated as a Braille trail, designed to attract the 15,000 blind residents in Orange and Seminole counties. There would be Braille markers along the path and pamphlets in Braille describing the plants at each marked station, where a blind or visually impaired person could, for example, smell a gardenia and then read about it. The club also intended to make an application to the Lions Clubs International Foundation in Oak Brook, Illinois, to fund an additional $10,000 toward the $66,000 required for completion of the trails and boardwalk. A step toward raising some of the money was a

Sounds of Mead benefit concert by the Florida Symphony Orchestra in April 1992. Concert selections included "The Syncopated Clock," the overture to *The Marriage of Figaro*, and selections from *Oklahoma*. The event in the amphitheater recalled the occasion in 1967 when the Florida Symphony Orchestra performed its first outdoor night-time concert.

The graphic artist Al Carroll, responsible for signage work for Church Street Station, Universal Studios, and for Winter Park's College Quarter, created a new logo and designed a new brochure for the Garden, both essential components in any rejuvenation project. His inspiration for the logo came from "the cycle of water, sky, birds above, fish below, children sitting beneath a beautiful Cypress tree, and a display of morning glories, which grow profusely in the Garden."

8.1: Mead Botanical Garden brochures no longer feature orchids. Nevertheless, Al Carroll's effort from the 1990s (extreme right) struck a sensitive John Muir-type evocation of the Garden.

As time went by, the Mead Garden Preservation Association had another think about fund-raising, and scaled back its plan to raise $1.5 million for park improvements to a more realistic budget of $750,000. Given that lack of funds had always been the stumbling block to previous improvement efforts, it wasn't entirely clear why the community would support the fund-raising drive even to this level, but the drive continued with the initial objective of funding a boardwalk and Braille trail.

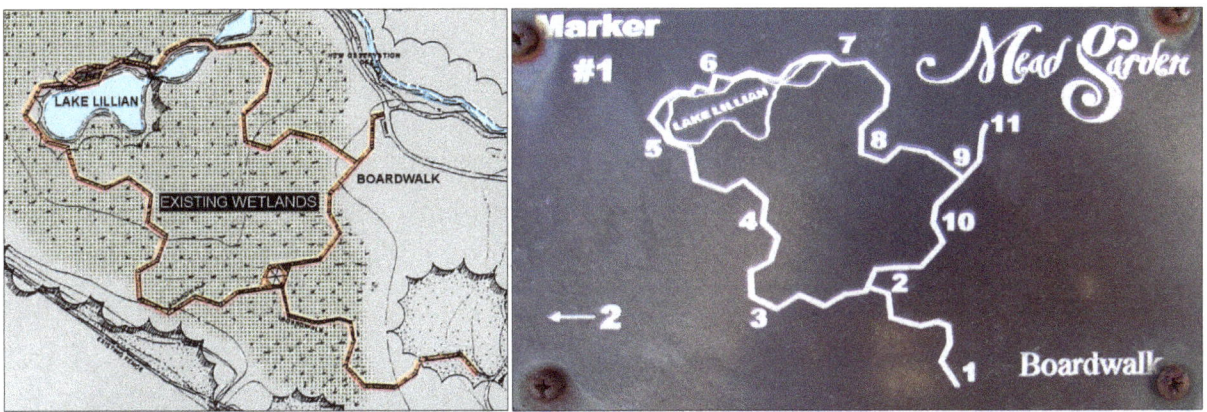

8.2: In 1993, the boardwalk element of the 1989 master plan that looped through the wetlands and visited Lake Lillian received financial backing (top left). Various stations were identified on the boardwalk where signs could be posted identifying objects of interest (top right). Under adult supervision, youth workers from Rollins College and the Florida YCC built the 2,500-foot boardwalk though deep, swampy muck (left).

In 1993, the Mead Botanical Garden Preservation Association, with additional funding from the City of Winter Park, the Orange County School Board, the Winter Park Lions Club and Rollins College, had sufficient funds to begin the 2,500-foot boardwalk. Before that, in the fall of 1992, environmental planning students from Rollins College had started to map out the course of the boardwalk to identify the best locations for visitors to learn about nature through their senses. They tramped their way through the mud, identifying ecosystems and looking for sites of educational interest where signs could be posted. The signs would offer park visitors tips on spotting wildlife or noticing changes in water levels. Other stations for the visually impaired would encourage visitors to feel the difference in bark textures or leaves or read about organic decomposition while breathing the wetlands' odorous air. A map of the proposed boardwalk and the stations identified by the students as signage points was delivered to city officials.

Under adult supervision, a team of youth workers from the Florida Youth Conservation Corps built the boardwalk through the wetlands. The task was made more difficult by the deep swampy muck that offered no firm footing and in places was up to 45 feet deep. Anecdotally, the story went that while working one day, someone heard a lot of screaming, and on investigating discovered and rescued a worker buried almost up to his neck in the mud.

On Saturday, October 2, 1993, the City of Winter Park dedicated the new boardwalk at Mead Garden at an event attended by county and city officials. The Killarney Elementary School Chorus opened and closed the ceremony with several songs, inviting the audience to sing along on some of the selections. Following the dedication ceremony and ribbon cutting, guests were invited to stroll along the newly completed walkway. They enjoyed the opportunity to get a closer look at native plants, trees, and wildlife from the convenience of the boardwalk, and were interested in the various signs and pictures explaining the flowers and wildlife that could be discovered there. The guests were told that the Braille wooden plaques would soon be in place along the trail, and then flyers written in Braille describing a history of the Garden would be given to visually impaired visitors. The flyers characterized the run-down Garden, without a trace of irony, as "The creation of beauty is man's greatest gift to those who come after."

In 2002, the Landscape Division of the Parks and Recreation Department worked with local senior Girl Scouts on improvements to the trail, planting fragrant plants and those with unusual textures, and attaching wind chimes for sound enhancement.

In late 1992, what was lying in wait for the Garden had nothing to do with beauty, nor its creation. One newspaper described this gift from Nature as, "Alien vine attacks tranquil park." Fueled by rainy weather, invasive skunk vine (*Paederia foetida*) had gained a significant toe-hold in the Garden and was threatening to take over increasing sections of the tree canopy. As usual, with the only criterion being minimal maintenance at minimum cost, the city was unaware of the intrusion until alerted to the problem by a letter to the Mayor of Winter Park, David Johnston, and Parks and Recreation Director, William Carrico, dated September 30, 1992, from two concerned members of the Winter Park Garden Club. The letter came from Barbara Roberts and Ruth Leonard, who described a planned trek through the Garden and their reaction when they entered an area where the trees were smothered with vines. "We stood transfixed and shocked by what we saw," they said, and pleaded with the city to do something about the fast-growing creeper, or in time, they argued, Mead Garden along with the surrounding area would be covered and destroyed by the vine's rampage.

Gordon Greger, city horticulturist, investigated and confirmed the plant's identity. He said that he had first noticed it a few years ago but that in recent months it had grown extensively and aggressively, claiming residence on up to six or so acres in the north-east wetlands part of the Garden. "It looks like something out of a science fiction movie the way it drapes over the trees," he said. With a growth rate of two to three inches a day, the vine was easily capable of overrunning everything in its path, including sixty-foot-tall magnolia trees.

Horticultural advice varied from waiting for a freeze to kill it off, together with the thousands of seeds the vine produced, to digging it out or cutting it down, actions that would be labor intensive and might spread the seeds even further. An alternative would be spraying the vines with chemicals that would be effective but expensive, and could affect adjacent desirable plants. As the debate about what to do continued, the vine kept on growing, blanketing out the sunlight from reaching the trees and everything else in the undergrowth.

8.3: In 1993 (top left) and 1994 (right), invasive vines swamped the Garden's tree canopy. Gordon Gregor, City of Winter Park's horticulturist, identified the major culprit as skunk vine (bottom left).

This infestation was the first warning, and the skunk vine was eventually joined by another invasive climber – the air potato vine (*Dioscorea bulbifera*). It was two invasive vines versus one city worker and local volunteers. Some yanking, cutting and pulling by dedicated volunteers took place throughout the following few years, but mostly the two vines had the Garden all to themselves and flourished.

In 1995 there was an urgent call for volunteers to pull weeds as things were getting out of hand. In the summer of 1997, the biennial meeting of the Mennonites took place in Orlando, at the Orange County Convention Center, attended by around 6,000 people and including approximately 1,000 teenagers. A hallmark of the Mennonite faith is service to the community, and after religious activities in the morning, the youngsters were organized in teams for afternoon service projects across Central Florida that included things like working in soup kitchens and food banks, as well as outdoor activities. One activity chosen by the teenagers was the pulling of skunk vine at Mead Botanical Garden.

For four consecutive days in the afternoons of late July 1997, teams of 250 youths began a concerted effort to rid the worse areas of the Garden of skunk and air potato vines. Before putting on the gloves and insect spray, the group was briefed about Florida's heat, insects and reptiles and shown how to identify the unwanted climbers that were smothering trees, choking off azalea bushes and snaking around the palmettos. With this effort by one of the largest volunteer group ever seen in the Garden, the growth of the vines was temporarily checked.

8.4: *In the summer of 1997, the biennial meeting of the Mennonites took place in Orlando, and weed-pulling at Mead Botanical Garden was an activity selected for community work involving teams of 250 young people.*

Of course, the vines soon came back and thicker than ever. In 1998, the City of Winter Park launched a major effort to rid the Garden of the foul-smelling climber, by severing the vine stems, cutting off the nutrients so the plant would wither and die. Since the plant propagates vegetatively, this act of pruning is of some help but frequently results in new plantlets being created from the existing rootball, or stem fragments, and can result in an even more aggressive growth pattern. "Wherever the stem touches down it will root and start a brand-new plant," warned Gregor. Carefully digging up the plant and removing the rootball was the only surefire way of ridding the area without the use of a targeted herbicide.

Slowing the invasion was one thing, but eradication was far more difficult. In 1999, the city switched to the use of herbicide in the battle to control the vines. Once the weather became warmer, and the vines were growing, herbicide spraying began using Garlon™ 4 at a concentration of ¾ wt.%. Initial control was good, but regrowth from the abundant seed bank required a second or third retreatment. Inevitably some good plants were affected, but according to Cory Clarke, deputy director of the parks department, "At the moment, it is the lesser of two evils."

Year	Degree of Infestation	Herbicide & Rate	Area Treated (acres)	Initial Control	Re-treatment
1999	Heavy	Garlon 4	4.0	95%	2000, 2002, 2005
1999	Light	Garlon 3A	2.0	85%	2000, 2002, 2005
2000	Heavy	Garlon 4	1.5	80%	2000, 2002, 2005
2001	Heavy	Garlon 4	3.0	95%	2002, 2005
2001	Light	Garlon 3A	1.5	90%	2002, 2005

8.5: Data from the City of Winter Park showing initial herbicide and follow-up treatments.

Having a dense canopy of vine material covering and linking trees together had another devastating effect in high winds when it would act as a sail and topple trees like dominoes. In the middle of August 2004, the path of a category four hurricane named Charley traversed the Orlando area of Central Florida from south-west to north-east. Hurricane-force winds and rain battered the tree canopy and brought down many prized trees, together with the vines in the Garden. About 10–15% of the bay canopy was lost to the storm, with the trees uprooted and branches broken, but the vines just continued to grow.

By 2005, the vines were on the verge of overwhelming the Garden again, and it was time to call in the professionals. The Florida Department of Environmental Protection operated an Upland Invasive Exotic Plant Management Program, and the East Central Regional Working Group was responsible for infestations in Orange County, where Mead Garden is situated. The Working Group chose Mead Garden as one of their control projects for that year, with a budget allocation of $10,820.

8.6: *When accurately surveyed, the Florida Department of Environmental Protection identified three category one invasive plants in the Garden; skunk vine, air potato, and wild taro (left to right).*

The initial survey showed that about a third of the Garden was affected, with the infestations being most extensive on the north side around the bay swamp, where there was a ninety percent cover of invasive plants. On the east side along the creek sixty percent of the area was affected. The western and southern boundaries of the Garden were light to moderately infested, with a thirty to sixty percent range of coverage, primarily along the garden borders. Three category one invasive plants were documented; the known skunk vine and air potato, but lurking in the undergrowth, particularly in the wetlands area, were significant infestations of wild taro (*Colocasia esculenta*). The Florida government funded control of the twelve most heavily infested acres, with the City of Winter Park providing in-kind services valued at $10,045 to control ten moderately infested acres adjacent to Howell Creek. The herbicides Garlon™ 3A and 4 were used again for the skunk vine and Roundup® for the air potato and wild taro, all applied as sprayed foliar treatments.

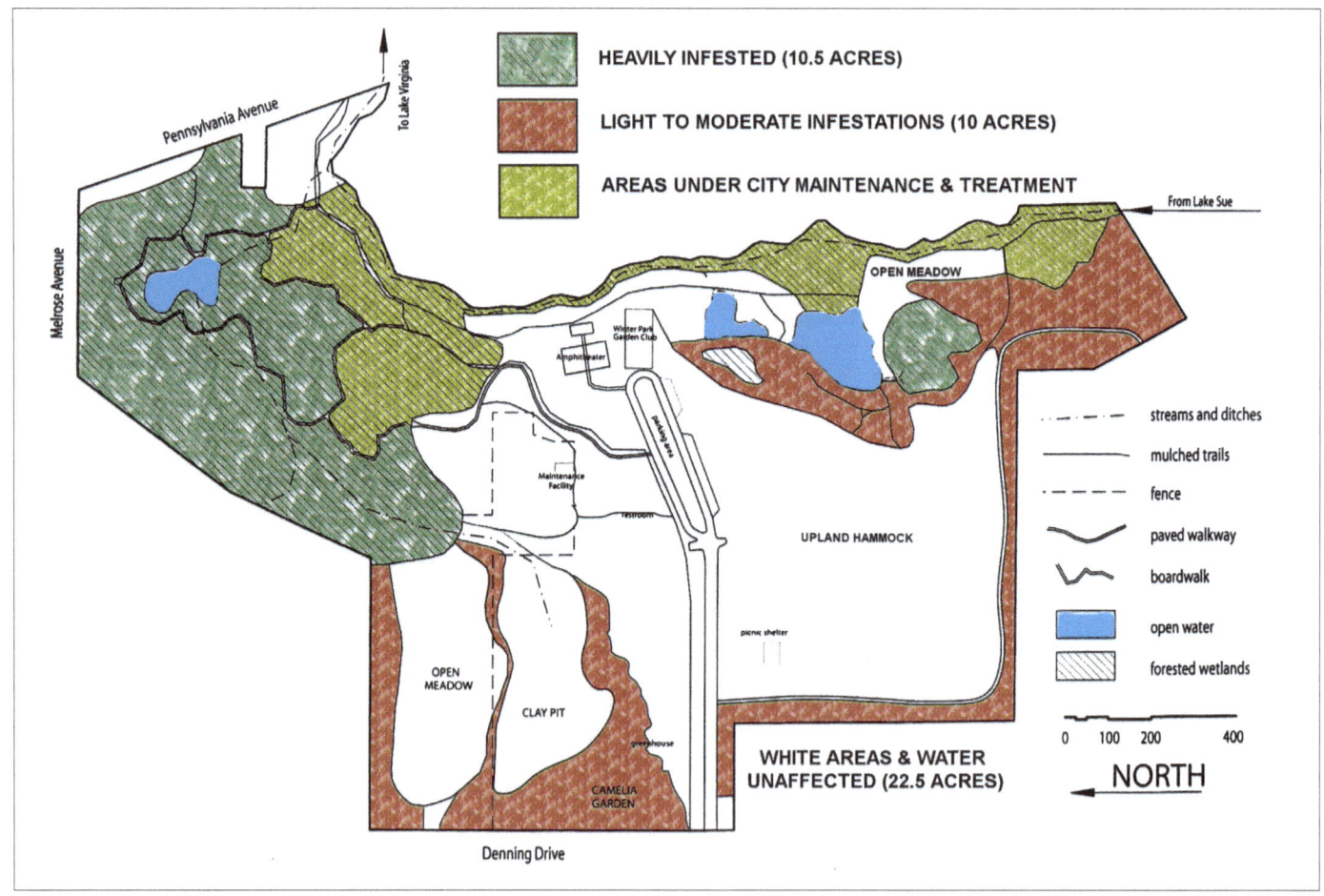

8.7: *There was extensive infestation of invasive plants in Mead Botanical Garden around the turn of the century. Almost all areas were affected with the exception of the pine uplands.*

By the start of the 21st century, the long-drawn-out process of neglect and decay had changed a once world-class botanical garden and tourist attraction into just another weed-covered community park that could be maintained at a subsistence level by one or two city employees. For years the city had tried to rename it "Mead Garden," dropping the "Botanical" so no-one would think it was one; confusing everyone by adding the occasional "s" on the end of the word "Garden"; and officially classifying it as a community park.

The destruction of the riches and horticultural beauty had begun in the mid-1970s with the removal of almost every flowering plant from the Garden. All that remained of the original plantings were the camellias, which had been relocated in the late 1960s from their preferred acidic-peat environment to the sandy-soiled west of the Garden, a lone cycad and stand of bamboo, a few night-blooming cereus, and some of the original palm trees planted by Jack Connery. Overall, the result had been the elimination of plants with a Mead association, including all of the orchids, and an overall beauty score of one out of ten. For winter visitors expecting a botanical garden and finding a neglected community park, it was a definite letdown, and as a memorial to Theodore L. Mead, a sustained insult.

However, for many local people, bird-watchers and dog-walkers, it was an ideal community park, and there was, and still is, a strong vocal minority bent on keeping it as natural as possible and avoiding what many of this group refer to as a "manicured look." But Mead Botanical Garden never was manicured; its beauty was the managed combination of informal flower beds interspersed with large areas left in their original but groomed natural state, crisscrossed with peaceful sunlit trails and the sounds of the cascading creek.

Since it was incorporated in 1937 as a non-profit enterprise, the one thing the Garden never lacked was grandiose and expensive conceptual development schemes. It was no surprise then when, in 2006, there was a call for yet another master plan to join the 1937, 1967, and 1989 unfulfilled plans collecting dust in someone's filing cabinet. The city engaged the design firm of Post, Buckley,

Schuh & Jernigan (PBS&J) to research and prepare the plan from which a strategic direction for the development of the Garden for 2009–2014 would be forged. Providing oversight to this effort were members from The Friends of Mead Garden, The Winter Park Garden Club, The Florida Federation of Garden Clubs, Orange County Environmental Education Program, the Winter Park community, the Winter Park City Commission, and local landscape architect professionals. A total of 13 public meetings were held with stakeholders, task forces and the steering committee and the completed plan, which cost $103,479, was presented to the City Commission on April 23, 2007.

The plan drew on many of the 1989 master plan features but importantly included a Mead Heritage Garden to display essential plants from Theodore Mead's collection, and a Mead Heritage House. Among the new gardens proposed were a butterfly garden, a discovery garden, and wildflower meadow with upland, marsh and wetland areas. It called for a rerouting of the brick entrance drive and vehicular traffic around the Garden; the relocation of the beloved 1950s amphitheater; the creation of an environmental and visitor center; and a new pedestrian entrance at the corner of South Pennsylvania and Melrose Avenues. The reintroduction of historical collections and gardens was a specific element of the city's strategic plan for the preservation of Mead Botanical Garden, but right at the end of its priority list and to be implemented only via the Friends of Mead Garden fundraising efforts.

Despite the 2007 master plan, and the fine words and high expectations of the 2009–2014 strategic plan, the Garden continues to present a somewhat neglected face to visitors, with few of the aspects of the grand plan translated into practical changes. Clearly, there is no point in having a master plan and a grand vision if the resources are not forthcoming to implement it, and the lessons of the past should have indicated this as the most likely outcome. As Winston Churchill once said, paraphrasing George Santayana, "Those who fail to learn from history are condemned to repeat it."

8.8: The 2007 master plan. 1 = main entrance, 2 = pedestrian entrance, 3 = visitor center, 4 = city maintenance yard, 5 = horticultural department, 6 = Mead heritage house, 7 = pavilion, 8 = amphitheater, 9 = marsh ecosystem, 10 = upland ecosystem, 11 = clay pits, 12 = camellia garden, 13 = discovery garden, 14 = butterfly garden, 15 = wildflower meadow, 16 = Winter Park Garden Club, 17 = restored wetland, 18 = retention pond, 19 = brick entrance drive, 20 = parking, 21 = bike trail, 22 = elevated boardwalk, 23 = at grade boardwalk, 24 = nature trail, 25 = learning pavilion.

Since the City's priority covered only essential maintenance appropriate to a community park, bringing back any of the original botanical nature of the Garden and connection to Theodore Mead could only happen through the effort and energy of volunteers, preferably with a leader of ability and vision. In 2012, the Garden was fortunate to attract such a person in the form of horticulturist Randal Lee Knight – Randy to his friends.

8.9: Left: Randy Knight advises volunteer Georgene Parsons on some of the finer points of gardening in the newly created Mead Legacy Garden of 2012. Right: The greenhouse in 2010 before it received its new roof.

With a lifetime of horticultural experience and knowledge relevant to plant growing in Florida, Randy together with a posse of faithful volunteers set about transforming the garden in and around the dilapidated 1978 greenhouse. With Randy's vision, the greenhouse received a new roof and a Mead legacy garden was created to join the already successful butterfly garden, established earlier by volunteers Ann Clement and Alice Mikkleson. Since then, and through the efforts of many volunteers, more areas of the garden have been replanted with shrubs and flowers, and orchids and other semi-tropical plants are once again back in the greenhouse. The camellia garden to the west of the property has been significantly expanded over the last ten years and now has a world-class collection of over 500 beautiful flowering specimens to delight visitors. And the night-blooming cereus, originally from the Mead collection, astounds all who see it on the one night of the year when it flowers. Currently, areas of the Garden dedicated to plants in the cycad, fern and daylily family are in the planning stage, driven by untiring and dedicated volunteers. Also, the Central Florida Camellia and Hibiscus societies continue to run shows every year, being true to their historical connections with the Garden.

Mead Botanical Garden has been likened to an unwelcome rich little orphan left as a financial liability on the doorstep of the City of Winter Park, and for the bulk of its upbringing with the city it has been treated as such. But hope does spring eternal; changes are happening to the Garden and areas of horticultural beauty are being slowly recreated by determined volunteers. The fortunes of the Garden are now on a more positive trajectory, and the survival of the Garden after all these years as a peaceful oasis of calm is something to be thankful for. As we nudge closer to the end of this decade, the unwanted step-child once wearing beautiful garments but now a little raggy could finally don some new clothes, achieve more positive community recognition as a garden not just a park, and become a fitting legacy to the memory of Theodore Mead.

Acknowledgments

I start by thanking the two key people who helped so much with the first book; the staff of the archives of Rollins College, Winter Park, Wenxian Zhang and Darla Moore, who could not have been more helpful in providing access and encouragement for me to finish this companion volume to Theodore Mead's biography. In addition to the T. L. Mead collection, the archives of Rollins College have recently acquired the minutes and records of the Winter Park Garden Club, who occupied a building in Mead Botanical Garden and as a result were in a unique position to capture many early photographs of the development of the Garden and add them to their scrapbook collection. Many of the images appear in this book and I gratefully acknowledge the club members who had the foresight to record and preserve them.

Central to the full historical account of the Garden were the articles and photographs published in archival versions of the Orlando Sentinel group of newspapers. These form the primary backbone of the book and are the main source of research material spanning the years. It is no exaggeration to say that a volume of local history like this could not have been written without this information, and the effort involved in making these digitized accounts available is sincerely recognized.

My warmest thanks go to Edwin Connery, his wife Nancy and their daughter

Cindy, for their kind hospitality and interest in getting the story of Jack and Helen's contribution to Mead Botanical Garden correctly documented. The icing on the cake when I visited them in Cape Coral was to see the outstanding portrait of Theodore Mead done by Sam Stoltz that hangs in their hallway.

The author (left) with Edwin Connery in front of the Sam Stoltz portrait of Theodore Mead.

I also wish to acknowledge the assistance of Rachel Simmons, archivist of the Winter Park Public Library for invaluable help; Madalyn Murphy of the Orange County Regional History Center for supplying one of the photographs; Joanne Wojtyto, archivist of the First Unitarian Church of Orlando for the picture of Eugene Shippen; Cindy Bonham, City Clerk, for searching the public records at the City of Winter Park; and conversations with Cynthia and Jim Hasenau, Beverly Lassiter, Nancy and Liz Davila, and Anne Sondag that provided useful perspectives and frequent additions to the text.

I was hugely fortunate to have two sterling proof-readers. The first, my youngest brother Peter Butler, ex-newspaper editor and hawk-eyed grammatical expert, picked up the typos and advised me on comma usage; the second, my good friend Mervyn Hathaway, scanned the text at lightning speed with his photographic memory and pointed out the inconsistencies. Both did a fantastic job, and in the unlikely event of any remaining errors, these remain my responsibility alone.

I am grateful to Jose at Pedernales Publishing, who performed a flawless book formatting, made my draft cover look professional, and put up with all my niggling little changes. I thank my youngest son, Elliott Butler, who sympathetically colored some of the original black and white images in Photoshop to bring them alive. And last, but certainly not least, an expression of loving gratitude goes to my long-suffering wife Jane, for living with Theodore Mead's ghost for the ten years or so "Orchids and Butterflies" and "Hope Springs Eternal" have taken to write.

Notes

Primary Source Abbreviations

RC-TLM	The Theodore L. Mead Collection, Archives & Special Collections, Rollins College, Winter Park, Florida. A collection of 31 boxes of letters and other printed material, some diaries and a few photographs.
RC-WPGC	The Winter Park Garden Club Collection, Archives & Special Collections, Rollins College, Winter Park, Florida. A collection of 12 boxes of yearbooks, reports, newsletters and other printed material, some loose photographs and scrapbook images.
WPPL	Articles and material related to the Mead Botanical Garden, Winter Park Public Library Archives, Winter Park, Florida.
WC	The Willis collection of private papers and photographs belonging to the Willis family, descendants of Edith Mead, housed in the ancestral Edwards home at Coalburg, West Virginia.
ECC	The Edwin Connery and family collection. Personal communication with the son of Jack Connery and access to his collection of private papers and photographs at his home in Cape Coral, Florida. Additional information from Edwin's daughter, Cindy, living in Windermere, Florida.

PROLOGUE

John Hurd Connery (1908–1982) preferred "Jack" as his Christian name according to his youngest son, Edwin, and hence is referred to as such throughout the book (ECC).

It is not known how the original commonly quoted figure of 55 for the acreage of the Garden was calculated. The 2018 GIS figure is 47.61 acres for the Garden and 0.41 acres for the Leedy plot, making 48.02 acres as the total land area (https://maps.ocpafl.org/webmapjs). Possible sources of discrepancy might be the inaccuracy of historical measurement and also the position of the east bank of Howell Creek which appears to have been altered over the years with the elimination of the original large and wide mirror pools and the development of the Sevilla subdivision.

Details of the opening ceremony come from "Hundreds Stroll Thru Mead Garden on Opening Day," *Orlando Sentinel*, January 15, 1940. The formal piece of land where the tulips were planted is believed to be close to the sloping ground where the old amphitheater is now, based on a 1940 aerial photograph. Connery's palm planting feat appears in "The History of Mead Botanical Garden," an undated but probably early 1960s manuscript by Grover in RC-TLM.

The propagating greenhouse by the Winter Park entrance contained many cuttings ("Mead Garden Draw Many Local Visitors," *Orlando Sentinel*, May 29, 1938) and is illustrated in the *Orlando Sentinel*, January 11, 1940.

CHAPTER ONE

The meeting between Jack Connery and Theodore Mead in 1922 at the Silver Lake scout camp and Jack's promise to build him a memorial garden appear in the article by Edwin Osgood Grover, "The Making of a Botanical Garden," *Parks & Recreation* (1948): 451-456, RC-TLM. Silver Lake and Camp Wewa in Plymouth, Florida, were popular sites for scout meetings and Mead attended gatherings there throughout the 1920s, as recorded in Mead's diaries (WC). Mead's party piece, the song *Taranty my son*, is referenced in a letter from McVoy to Mead, May 6, 1901, RC-TLM.

Connery joined the Beebe and Barton deep-sea diving expedition in 1930 initially as a cook (ECC, personal communication) but together with Robert Whitelaw was named as official photographer in the published records of the New York Zoological Society (William Beebe, "Bermuda Oceanographic Expeditions 1929-1930", *Zoologia*, Xlll, no. 1 (1931): 12. Connery photographed many aspects of the 1930 expedition, and the book by William Beebe, *Half Mile Down* (New York: Harcourt, Brace & Co., 1934), references Connery as the photographer for figures 31 to 44. Other references to Connery's part can be found in Brad Matsen, *Descent: The Heroic Discovery of the Abyss* (New York: Knopf Doubleday Publishing Group, 2007), on pages 69-70, 73-75, 79-80, and 87-88. The report of Connery's injury appears on page 102. The June 1931 issue of the *National Geographic* entitled "A Round Trip to Davy Jones' Locker" is illustrated with photographs by Jack Connery (https://openexplorer.nationalgeographic.com/expedition/bathysphere).

Connery's agreement with Rollins College is stated in Tucker Loane Farrell, "Dr. Grover Tells His Tale," *Orlando Evening Star*, June 30, 1961. His many visits to Mead's estate are documented in Mead's diaries from 1930-1934 (WC). In 1932, for example, he recorded Connery's visits as April 15 (brought Rollins students), April 18 (projector entertainment to scouts on birds), and July 22 (digging caladium and repotting orchids). The trip to Daytona Beach is reported in a letter from Mead to Ogden Willis, July 5, 1932, WC. His birthday party surprise at the Connery's in 1933 is described in a letter from Mead to the Willis family, February 24, 1933, WC.

It is likely that Grover visited Mead's garden on many occasions, but records of just two of these visits appear in Mead's diary of December 12, 1926 and August 28, 1927, WC. It is likely, but not recorded that Edwin's brother Frederick was also present on one or both of these visits. "I want to meet the famous Mr. Mead" comes from Farrell, "Dr. Grover Tells His Tale."

The confusion with Mead's will and the orchid ownership comes from Paul Butler, *Orchids and Butterflies*, (Winter Park, Little Red Hen Press, 2016), 287.

The suggestion that Grover's brother might be the director of Mead's estate if it were given to Rollins College appears in a letter from Grover to Willis, January 6, 1937, RC-TLM. In December 1937, the tennis courts on the north shore of Lake Eola were torn down and rebuilt at the high school on East Robinson Avenue (Eve Bacon, *Orlando a Centennial History, Volume II*, The Mickler House, Publishers, Chuluota, Florida, 1977, page 86).

The circumstances related to the first meeting of Connery and Grover in early 1937 and the proposed location of the Garden are recounted in a Grover-written manuscript (believed to be unpublished) entitled "The History of Mead Botanical Garden," dated probably early 1960s, RC-TLM. Realtor Walter Rose donated property for the development of Mead Garden ("Walt Rose Played a Vital Role in the Development of the Area," *Orlando Sentinel*, April 7, 1988); was a promoter of beautification in his subdivisions ("Rose Works for Beauty," *Orlando Sentinel*, June 13, 1937); and could see the business self-interest in donating property ("Walter Rose, Outstanding Citizen, Developer, Dies," *Orlando Sentinel*, September 1, 1958). The actual area donated was 18.92 acres according to the deed. Grover's comment "best afternoon's work" comes from "Dr. Grover Tells His Tale," *Orlando Evening Star*, June 30, 1961.

A copy of the "Charter of the Theodore L. Mead Botanical Garden" dated May 11, 1937, Incorporation Book 7, Page 147, is contained at RC-TLM. The Orange County Comptroller website (http://or.occompt.com/recorder/eagleweb/docSearch.jsp) displays the Rose deed as Deed Book 524, Page 413. This original deed was later rewritten (Book 530, Page 359) and then amended with a release of certain restrictions (Book 541, Page 438). Lillian Treat Simkus was born on July 31, 1932, and was the daughter of Clyde and Anna Simkus (née Treat), and the lake was formally named "Lake Lillian" to conform to one of the deed conditions (James A. Treat, June 7, 1939, Deed Book 537, Page 112).

The Bartels deed can be found in Book 520, Page 63, and the Orange County ownership of the clay pit return to the City of Winter Park is recorded in Book 537, Page 115. Note that the 1.96 acres of what used to be called the "Clay

Pit," is not the same as the larger area which is now called by that name. The Ruth Scott Leedy and R. F. Leedy deed can be found in Book 499, Page 201; the Cherokee Park plat is dated November 17, 1925, Plat number 1925P00L137.

For 30 years, there was an expectation that the land to the east of Howell Creek would one day be part of Mead Botanical Garden. The first vision of the Garden of 1937 showed this, as did the 1967 master plan. In 1938, the City of Winter Park bought lots 1&2 and 6&7 of Cherokee Park at auction, paying delinquent property taxes (Book 510, Page 67). In 1943, Grover bought lot 5 (Book 602, Page 412), selling it to the city in 1956 (Book 104, Page 38). The city bought lot 4 in 1957 (Book 245, Page 249), but unfortunately never acquired lot 3 of 0.15 acres, which to this day remains an unfortunate protrusion into the Garden. Possession of the entire triangular tract would have allowed the creation of a formal Winter Park entrance to the Garden for residents and Rollins students.

Grover's presentation to the City Commission is reported as "City Botanical Garden Plans are Approved," *Winter Park Herald*, April 9, 1937, RC-TLM. In the Great Depression, a time of years of tight money, bread lines and thousands of jobless, the Works Progress Administration (WPA; renamed in 1939 as the Work Projects Administration) was the largest and most ambitious of the American New Deal agencies, employing millions of people (mostly unskilled men) to carry out public works projects. It was a national program that operated its projects in cooperation with state and local governments, which provided a share, generally 10–50% of the costs. Once approved, Mayor Moody called for immediate action ("Memorial Garden to Dr. Mead Planned," *Orlando Sentinel*, April 11, 1937).

Martin Andersen's reaction to Mead Botanical Garden as "Who in the devil is in back of this?" comes from "Dr. Grover Tells his Tale," *Orlando Sentinel*, June 30, 1961. The three-column piece in praise of the Garden appears in "Memorial Garden to Dr. Mead Planned," *Orlando Sentinel*, April 11, 1937. Robert Michell was an inspired choice to become an associate of the Garden with his family connection to Henry Nehrling (https://floridahorticulturehistory.wordpress.com/tag/robert-mitchell). The plan proposed that once the land to the east of Howell Creek was acquired, an imposing entrance would be created off S. Pennsylvania Avenue leading to the aquarium ("John Connery Started Winter Park Movement," *Orlando Sentinel*, November 7, 1937).

The educational benefits of a botanical garden were stressed in the original WPA grant application ("$200,000 Beauty Spot Planned in Winter Park," *Orlando Sentinel*, October 31, 1937). "Mead Project Funds Given Out," *Orlando Sentinel*, December 19, 1937, announced the success of the application, and contained excerpts from various congratulatory letters. The indication of a willingness to add to the grant appears in "WPA Grant," *Orlando Sentinel*, January 7, 1938.

News of the groundbreaking was announced as "Work Starts Tomorrow on the Mead Botanical Garden," *Orlando Evening Star*, January 7, 1938. Reports of the event were published as "Mead Garden Started Today," *Orlando Sentinel*, January 8, 1938, and "Ground Broken for Mead Park," *Orlando Evening Star*, January 9, 1938. Two photographs of the groundbreaking appear to have survived; in the first, shown here, nine dignitaries (L to R: Way, Barbour, Jackson, Grover, Leedy, Apgar, Connery, Thompson, Moody) stand behind Senator Andrews. In the second picture, reproduced in the *Orlando Evening Star*, Andrews has moved slightly to the side to allow Senator Walter Rose to join the group, unfortunately blocking the figure of Barbour apart from his pant leg.

The first trail in the Garden was quickly completed ("Botanical Unit to Open Soon," *Orlando Sentinel*, February 17, 1938) and other trails and landfill improved with sand and clay supplied by the borrowed dragline from the county ("Dr. Grover Tells his Tale," *Orlando Sentinel*, June 30, 1961). Connery's progress report to Senator Andrews is contained in a letter dated April 16, 1938 (RC-TLM).

The amount of the WPA grant was explained in "Why WPA Funds Given to Mead Botanical Garden Were So Low," *Orlando Sentinel*, April 27, 1938, and this reference also mentions the $3,000 that Connery put into the Garden from his own pocket. The restrictions on WPA hours worked, and the lack of money for materials and supplies appear in "Mead Garden Draw Many Local Visitors," *Orlando Sentinel*, May 29, 1938. The investment conditions of the "Revenue

Certificates" were detailed in "Facts About the Mead Garden," *Orlando Sentinel*, November 5, 1939. Many local organizations supported the Garden with materials and supplies, either via generous donations or by the use of revenue certificates ("Botanical Garden Fitting Memorial to Mead," *Orlando Sentinel*, January 2, 1938). Martin Andersen was an admirer of the revenue certificate approach ("How They Built Mead," *Orlando Sentinel*, September 19, 1939). A description of all the work achieved to the end of May 1938 is given in "Mead Garden Draw Many Local Visitors," *Orlando Sentinel*, May 29, 1938. The application for a supplement to the original WPA grant is recorded there as well as in "Mead Garden Ask WPA Grant," *Orlando Sentinel*, July 1, 1938.

The difficulties that the WPA had with the wording of the Rose deed were initially defended by Rose (Letter Rose to Dill, November 18, 1938, RC-TLM), but eventually there was a rewrite, as detailed on page 196. Treat's lack of a written deed triggered a Connery/Grover letter to Mayor Moody (draft dated May 16, 1939, RC-TLM), and a further letter to Moody (Grover to Moody, June 3, 1939, RC-TLM) expressing frustration.

Newald's supply of the truck against a revenue certificate is in a letter from Helen Connery to Mrs. Sprague-Smith, dated May 28, 1939, RC-TLM. Newald's reply when thanked appears in "How They Built Mead," *Orlando Sentinel*, September 19, 1939. Hauling in palms from Oviedo is described in a letter from Connery to Eugene Shippen, dated June 25, 1939, RC-TLM, and the trip down to the Everglades in "Mead Garden to be Opened in January," *Orlando Sentinel*, June 11, 1939.

Grover's attendance at the World's Fair is noted in Eduard Gfeller, *Edwin Osgood Grover: The Business of Making Good* (Winter Park: Eduard Gfeller, 2016), 38. Before getting the New York train from Jacksonville, he and the Connery's visited the Glen Saint Mary, the Gerbing Nurseries, and Bruno Alberts (letter from Helen Connery to Mrs. Sprague-Smith, dated May 28, 1939, RC-TLM). In 1937 Gustav George Gerbing transformed about seven acres of family property along the Amelia River into a public garden, Gerbing's Azalea Gardens, with plans for massive plantings of azaleas and camellias for viewing and sale. The area was expanded into fifteen acres and became Gerbing Gardens, with groves of native trees including cedars, magnolias, bay, pines, live oaks and water oaks draped with Spanish moss, and wisteria trees. Bruno Alberts was born in Louisville, Kentucky. In his youth, he was active as a landscape painter. After the death of his father, John Alberts Jr. in 1931, he moved his family to Mandarin, Florida where he developed a botanical business based on the growing orchids.

John A. Porter's death was reported in "Orlando the Loser," *Orlando Sentinel*, April 25, 1939. Eugene Shippen was born in Worcester, MA and came to Winter Park in 1930 where he met and became friends with his neighbor Irving Bacheller, and Hamilton Holt and Annie Russell ("Shippen Gives Longevity Rule," *Orlando Evening Star*, January 28, 1955). The Shippen home location and design by Rogers are given in *Orlando Evening Star*, March 6, 1931, and the description of the groundbreaking in "Noted Persons Present for Breaking of Ground for New Home," *Orlando Sentinel*, March 11, 1931. Grover got to know Shippen and invited him as a speaker at one of his events (*Orlando Evening Star*, April 14, 1932).

Robert Bruce Barbour arrived in Winter Park in 1932 and commissioned Rogers to design a home on the shore of Lake Osceola at 656 Interlachen Avenue. Barbour was a wealthy industrialist who retired from running The Eclipse Chemical Manufacturing Company of Chelmsford, MA, making indelible inks and aniline dyes (Fred Merriam, *Chelmsford Revisited*, Chelmsford Historical Society, Arcadia Publishing, p. 78). He had a love for azaleas and extensively planted them in his garden, and led a campaign to beautify the Winter Park station area with them ("Winter Park Wins Beautiful Park Thru Generosity and Vision of R. B. Barbour," *Orlando Sentinel*, July 26, 1936; "Beauteous Dream Becomes Reality," *Orlando Sentinel*, August 2, 1936; "Spain Comes to Florida," *Orlando Sentinel*, January 31, 1937). Holt's Spanish event was reported in "Winter Park News," *Orlando Sentinel*, May 30, 1939; the report of the beautification of Park Avenue in the *Orlando Sentinel*, April 13, 1939; and that of the tea in Mead Garden in "Mr. Bacheller Entertains at Tea," *Orlando Sentinel*, April 3, 1939.

Bacheller's parting words were quoted in "Irving Bacheller Leaves for North with Words for Mead Botanical Garden," *Orlando Sentinel*, May 2, 1939. Both Grover and Connery were full of praise for Shippen's generous donation, and sent letters of thanks described the work in the Garden in detail, including the waterfalls, gardenias, azaleas, camellias and palms (letter Grover to Shippen, undated probably August, 1939, RC-TLM; letter Connery to Shippen, September 8, 1939, RC-TLM). The letter from Connery to Shippen also described their visit to Heintzelman and the Bumby Hardware store. Progress in the Garden, including the sweet pea display, was recounted to Rose in a letter from Connery, dated October 4, 1939, RC-TLM. Gamble Roger's agreement to design one of the Garden entrances is contained in a letter he sent to Shippen, June 14, 1939, RC-TLM.

Connery's memo regarding the amount of work completed from July to October is recorded on an undated memo in RC-TLM. The fact that Helen Connery was Grover's personal secretary for a time comes from a letter she sent to Donna Rhein at the Winter Park Public Library on July 15, 1995 (RC-TLM). The occasion of Edwin Connery's christening with Grover as godfather and the choice of his Christian name was revealed at a personal meeting with Edwin Connery (ECC).

Andersen's editorial in the *Orlando Sentinel* of September 30, 1939, was entitled "Meaning of Mead Garden." His promise to publicize the Garden with photographs, special editions, and banner headlines appears in a letter Grover to Rose, October 4, 1939, RC-TLM. The dire financial state of the Garden was emphasized in "A Challenge to this Community," *Orlando Sentinel*, October 2, 1939. "Let's Get Going Orlando!" in the *Orlando Sentinel* of October 4, 1939, called for the setting up of a campaign, and a money drive was started ("Mead Garden Leaders Plan Money Drive," *Orlando Sentinel*, October 4, 1939). Grover's quote "Think what we can do with 55 acres" comes from "Mead Garden Drawing Power Told by Grover," *Orlando Sentinel*, October 10, 1939.

The spectacle of the Lake Lillian birds nesting around dusk appears in "Daylilies

Displayed," *Orlando Evening Star*, May 7, 1949. The number of people at the preview varied in the newspaper reports ("Mead Garden Preview Draws 1,000 to New Central Florida Beauty Spot," *Orlando Sentinel*, October 29, 1939; "5,000 Throng Mead Garden Yesterday," *Orlando Sentinel*, October 30, 1939).

The campaign to raise money for the opening started in earnest with "The Mead Garden Campaign Opens Today – Here is Why the Money is Needed," *Orlando Evening Star*, November 2, 1939. "Visit Orlando and See the Orchids," was on the front page of the *Orlando Sentinel*, October 31, 1939. The approximately two acres of the Garden at the very south, originally donated by Walter Rose, lay in the City of Orlando until annexed into Winter Park in 1955. The first day's total is taken from "Mead Fund Reaches $1,321 – Directors of Drive Satisfied," *Orlando Sentinel*, November 4, 1939. "Facts About the Garden," was a half-page block in the *Orlando Sentinel*, November 5, 1939, and the total by day eight appears in "$500 Given for Garden," *Orlando Evening Star*, November 10, 1939. The contribution of the Sentinel Group of newspapers to the campaign and the support of Martin Andersen, in particular, elicited a warm response from Grover (letter Grover to Andersen, November 15, 1939, RC-TLM).

Once the main orchid greenhouse was in place, many growers donated orchids to the collection ("Orchid Growers Present Plants to Mead Garden," *Orlando Sentinel*, November 19, 1939). Orchid photographs were featured in the local newspaper ("Our Orchid Garden New and Beautiful Sight for Visitors," *Orlando Sentinel*, November 26, 1939).

In 1940, Cypress Gardens was not yet a theme park, and water skiing had yet to be introduced. Dick Pope was known for his lively personality and his colorful clothing, including a turquoise suit trimmed in pink, worn with bright white shoes. Attendance figures for 1938 were given in a letter from Florida Cypress Gardens to Connery, August 16, 1938, RC-TLM, and the announcement of Pope as Publicity Director in "Dick Pope to Publicize Mead Garden without Pay," *Orlando Sentinel*, November 29, 1939.

Opening day was extensively covered in "Hundreds Stroll Thru Mead Garden on Opening Day," *Orlando Sentinel*, January 15, 1940. The arrival of bumper stickers comes from "First of Mead Garden Bumper Strips Arrive," *Orlando Sentinel*, January 10, 1940. Grover agreed with Andersen's promotion of Orlando as "The Orchid City" ("Florida's Orchid Cities," *Orlando Sentinel*, October 25, 1939, and letter Grover to Andersen, November 15, 1939) and a step in that direction was taken in February 1940 ("Orchids in Orlando Signs Placed all over State," *Orlando Sentinel*, February 16, 1940).

CHAPTER TWO

By March 1940, the Garden was a "great splash of color" according to the *Orlando Sentinel*, March 6, 1940. Reports of the azalea garden stressed the "Profusion of Blooms," *Orlando Sentinel*, March 7, 1940, and "2,000 Azalea Bushes at Mead Garden," *Orlando Sentinel*, March 11, 1940. Varieties of the Kurume-type are named in "Late Azaleas Displayed at Mead," *Orlando Sentinel*, March 25, 1940, and the Flame Azalea in an unpublished newspaper clip (presumably penned by Edwin Grover) in RC-TLM.

William Bartram described the 'Fiery Azalea' as the most brilliant flowering shrub known to date. He observed that giant clusters of flowers covered the bushes with incredible profusion along the sides of the rolling landscape, making it look like the hills were on fire (Mark Van Doren, ed., *Travels of William Bartram* (New York: Dover Publications, 1955), 264). It was the French botanist Andre Michaux, sent to the colonies in 1795 to collect plants for France, who received credit for the name Flame Azalea, *Rhododendron calendulaceum*, for this particular American spectacular flowering azalea instead of William Bartram, who had seen and identified several species and called them 'Fiery Azalea,' twenty-two years earlier than Michaux. It is not known whether Grover's report of the Flame Azalea in the Garden, which he described as having "almost a tangerine color," refers to *Rhododendron calendulaceum* or *Rhododendron austrinum*, the Florida Flame Azalea.

The second connection with Bartram in the Garden was the planting of a Franklin tree, reported as "Mead Garden Prepares New Plantings," *Orlando Sentinel*, February 12, 1939. This tree, a member of the tea-bush family, was first observed by John and William Bartram in 1765 in the neighborhood of the Alatamaha River, near Fort Barrington, Georgia, and rediscovered by young William Bartram in 1775 who named it "Franklinia Alatamaha" in honor of Benjamin Franklin. It is also referred to as the lost tree of the Alatamaha. The fate of this tree, planted near the Winter Park entrance, is unknown.

Grover was impressed by the many and varied tulip displays at the 1939 World's Fair in New York (www.1939nyworldsfair.com/postcards/CT-Art-Colortone/CTA-40.htm). Although the general opinion was that tulips could not be successfully grown in Florida, he pressed on with the experiment, recognizing the large amount of work involved (Letter from Grover to Mrs. Sprague-Smith, August 10, 1939, RC-TLM). The tulips arrived in November 1939 ("Tulips for Mead Garden Arrive from Holland," *Orlando Sentinel*, November 18, 1939), transferred to cold storage conditions (letter from Connery to Dr. and Mrs. Shippen, September 8, 1939, RC-TLM), and planted ("16,000 Tulips in Mead Garden for Experiment," *Orlando Sentinel*, February 15, 1940). By early March, the hillside was a mass of tulip color ("Tulip Experiment at Mead," *Orlando Sentinel*, March 10, 1940).

The report of Graham Grover's funeral appears as "Grover Funeral to be Held Today," *Orlando Sentinel*, March 6, 1940. The book on Grover by Gfeller, *The Business of Making Good*, describes the relationship between Graham Grover and Jack Connery (p. 83), considers the suicide of Grover's son a result of a schizophrenic disorder (p. 101), and postulates that as a result Grover sought solace in "his beloved Mead Botanical Garden" (p. 83).

The original reception lodge location was at the site of the current Garden Club building and was visible from the annual hillside garden, which was located at and above where the seating for the old amphitheater is now. Ground-breaking for the building was reported in "Work on Mead Botanical Garden Going

Forward Steadily," *Orlando Sentinel*, February 11, 1940. When opened, the shop sold postcards, Kodachrome film, and soft drinks, and became a natural display area for Harry Russell Ballinger's work ("Garden Paintings to be Exhibited," *Orlando Sentinel*, March 17, 1940). This artist chose Orlando for his home and Mead Botanical Garden as his subject ("Artist Chooses Orlando," *Orlando Sentinel*, February 11, 1940), presenting Grover one of his paintings ("Painting is Presented," *Orlando Sentinel*, March 17, 1940). Sam Stoltz's reception and the gift of his portrait of Theodore Mead is recorded in the *Orlando Sentinel*, February 2, 1941, "Mead Garden Holding Official Reception for Artist Sam Stoltz to Unveil Portrait."

The spring flush of flowers in the Garden is documented in "Mead Garden Greets Spring with Beauty," *Orlando Evening Star*, March 23, 1941, and "Roses, Amaryllis Make Mead Garden Mass of Color for Orlandoans," *Orlando Evening Star*, April 6, 1941. Daylilies hanging over Howell Creek comes from "Gardenias Blooming Now in Mead Garden," *Orlando Sentinel*, May 5, 1940. Jeweled caladiums and the heavy fragrance of the gardenias are mentioned in "Today is Orlando Day at Mead Garden," *Orlando Evening Star*, June 2, 1940.

Three main types of gardenias were present in the Garden. *Gardenia jasminodes* 'Belmont' (Cape Jasmine), a large evergreen shrub with semi-double ivory-white highly scented flowers; *Gardenia jasminoides* 'Veitchii' (Everblooming Gardenia), a medium-sized evergreen shrub with pure white flowers sweetly fragrant; and *Gardenia jasminoides* 'Hadley' (Hadley Gardenia), an evergreen compact hedge or corner foundation plant with flowers in creams and whites. By October the gardenias had grown so prolifically that they needed moving ("Mead Garden to Open Sunday Showing Many Rare New Flowers," *Orlando Evening Star*, October 20, 1940). Grover's story of how Connery obtained the rare gardenia comes from "Mead Garden Drawing Power Told by Grover," *Orlando Sentinel*, October 10, 1939.

Oil heating for the Porter greenhouse appears in "Work on Mead Botanical Garden Going Forward Steadily," *Orlando Sentinel*, February 11, 1940. The Bird of Paradise plant is mentioned in "Bird of Paradise Flower Blooming in Mead Garden," *Orlando Evening Star*, June 29, 1941. This plant belongs to the same family as the Traveler's Palm and the general shape is similar although smaller. George III's wife, Charlotte Sophia, of the Mecklenburg-Strelitz family, was a great patron of botany. When this stunningly-beautiful plant import was taken to England from South Africa, it was named for the Queen's family.

In the *Orlando Sentinel* of March 8, 1940, is a description of the Fiji Island Fern, and in the *Orlando Sentinel* of February 22, 1940, a report of the hormone treatment experiments in the smaller propagating greenhouse. Grover's comment about "new beauties continually from now on" and the 'Watch Us Grow' slogan come from "It's Mead Garden Today," *Orlando Sentinel*, January 14, 1940.

The list of requested roses from the Waxahachie Nursery, Texas, is in a letter Grover sent to Mr. Freeman, December 4, 1940, RC-TLM. The roses were delivered and planted on December 30, 1940 ("Huge Rose Shipment to be Planted Monday in Mead Garden Beds," *Orlando Evening Star*, December 29, 1940), and many were in bloom the following April. The raised and enlarged formal annual garden near the Orlando entrance is commented on in "Mead Garden Makes Many Improvements," *Orlando Evening Star*, October 19, 1940, and the work to alleviate flooding in "Mead Garden to Open Sunday Showing Many Rare New Flowers," *Orlando Evening Star*, October 20, 1940. News of the 1944 hurricane damage at the Orlando gate comes from "Mead Garden Little Hurt," *Orlando Evening Star*, October 28, 1944.

The Connery family moved into their new home at 1358 Richmond Road, Winter Park (*Orlando Sentinel*, September 23, 1941) with their two sons, John Hurd Connery Jr. (born in 1935) and William Edwin Connery (born in 1937). Their perilous financial state at that time was recalled by Edwin Connery (ECC) who also remembers Helen's hospitalization for malnutrition, referred to in Gfeller, *The Business of Making Good*, 83.

Work was started on the Orlando Air Base on the morning of March 18, 1941, under the supervision of Jack Connery ("No More Information on Airbase

Equipment," *Orlando Sentinel*, March 18, 1941, and in a little more than a year it had become a well-drained beauty spot ("Amazing Transformation Made in Orlando Air Base," *Orlando Sentinel*, April 6, 1942). Many more palm trees were planted to finish the work, as detailed in "16,000 More Palm Trees Set Out at Air Base," *Orlando Evening Star*, April 24, 1942. Florida posthumous Medal of Honor recipient Thomas B. McGuire, Jr. was among the pilots who trained there and later taught there. Major McGuire was the second highest scoring US ace of the war, achieved 38 victories in the Pacific, before being killed in action over the Philippines on January 7, 1945.

Competition for peat to camouflage the runways, and the comment by Connery to "take all you need," is taken from an unpublished manuscript written by Edwin Grover "The Making of the Mead Botanical Garden," RC-TLM. The manuscript is undated but talks about free admission to the Garden, so this places it in the early 1960s. Excavation of the two lily pools is described in Edwin Grover, "The Making of a Botanical Garden," *Parks & Recreation*, August 1948, RC-TLM, and in "Pond for Rare Lilies at Mead Garden Now Under Construction," *Orlando Evening Star*, August 8, 1942. The wide selection of tropical water-lilies of varying color were flowering by 1944 ("Truckload of Orchids Arrives for Display at Botanical Garden," *Orlando Sentinel*, November 19, 1944), to join the many annuals in bloom ("Work on Mead Garden Going Forward Steadily," *Orlando Sentinel*, February 11, 1940).

Grover's letter dated November 15, 1939, to Martin Anderson suggesting a Central Florida Flower Show can be found at RC-TLM. In early spring the amaryllis in the Garden were in full bloom ("Beautiful Amaryllis Blooming at Garden," *Orlando Sentinel*, April 5, 1942). The historical connection of this plant with Theodore Mead led to the establishment in 1940 of an annual amaryllis show ("Amaryllis Show to be Held at Mead Botanical Garden April 13 & 14," *Orlando Sentinel*, March 31, 1940). At this show, the rare blue amaryllis was on display; the only other plant known to be in flower was in Miami ("Blue Amaryllis is Now Blooming," *Orlando Sentinel*, March 31, 1940). This species was discovered growing in the granite crevices of cliffs on the steep mountainsides near Rio de Janeiro in 1860, but it was not until 1899 that it was introduced to gardens by the Englishman whose name it bears, Arthington Worsley (1861-1943).

The first annual daylily show soon followed ("Daylily Show at Mead to Open Next Sunday," *Orlando Evening Star*, May 12, 1940), with the accent on growing daylilies suitable for the Florida climate, as detailed in Wyndham Hayward, "The Daylily in Florida," Proceedings of the Florida State Horticultural Society, Vol 63, 194-198 (1950), but in the following year it was absorbed into the second annual amaryllis show ("Amaryllis Show to be Held at Mead Garden," *Orlando Evening Star*, March 16, 1941).

News of the first inter-state camellia show comes from "Inter-State Camellia Show Planned Jan 12," *Orlando Sentinel*, January 1, 1941, and "Flower Show Plans Ready," *Orlando Sentinel*, January 7, 1941. Hostesses were chosen ("Hostesses Named for Camellia Show," *Orlando Sentinel*, January 7, 1941) and the show was reported to be "quite a sensation" according to the *Orlando Evening Star*, January 15, 1941.

Descriptions of the Orlando Coliseum come from Joy Wallace Dickinson, "Great Depression Put Stamp on Lost Ivanhoe Landmarks," *Orlando Sentinel*, February 8, 2015, "Era of Big Bands Gone…So is W. R. Their Promoter," *Orlando Sentinel*, November 21, 1971, and "Time Machine's Bubbling for 1940s New Year's Eve," *Orlando Sentinel*, December 28, 2014. The show was enthusiastically endorsed by growers and florists ("First Florida Flower Show Being Planned for Orlando May 3-5," *Orlando Evening Star*, April 20, 1941) and opened by the president of the Florida Federation of Garden Clubs ("Ribbon is Cut to Open Giant Flower Show," *Orlando Evening Star*, May 4, 1941). The Garden Mart's display and Joe Leinhart Jr.'s prize-winning delphiniums are described in "Winter Park Winner of Blue-Ribbon Prize at State Flower Fete," *Miami News*, May 11, 1941. This newspaper edition also has the photograph of Helen Connery and the delphiniums. Joe Leinhart planted the first celery farm in Oviedo around 1898, and his son Joe Jr. (1893-1972) was one of Theodore Mead's protégés.

The Leinhart family were friends of the Meads in Oviedo, and Joe Jr. was a frequent visitor to Wait-a-Bit and later became a commercial grower and owner of the Oviedo Gardens nursery, primarily as a result of Mead's early horticultural influences.

Details of the 1942 Camellia Show and the camellia 'Alba Plena' were covered in "Magnificent Camellia Show to be Held at Mead Garden," *Orlando Sentinel*, January 11, 1942, and "Where Camellias Flourish," *Orlando Sentinel*, January 21, 1942. The 1947 show was described in the article "Mead Camellia Show Slated," *Orlando Evening Star*, January 21, 1947, and the 1950 best flower in the show (Glen40) in "Mead Garden Opens Annual Camellia Show," *Orlando Sentinel*, January 15, 1950. The show's success in popularizing the camellia comes from "Garden Set for Show," *Orlando Evening Star*, January 18, 1949, and consequently inter-state or Florida-wide shows moved to the larger Municipal Auditorium in Orlando (now the Bob Carr Theater).

The reports of camellia shows in 1951, 1952 and 1954 are: "Mead Garden to Feature Camellia Show," *Orlando Sentinel*, January 14, 1951; "Winter Park Camellia Show This Weekend," *Orlando Evening Star*, January 18, 1952; "Mead Garden Camellia Show Draws Throng," *Orlando Sentinel*, January 20, 1952; and "Camellia Girls," *Orlando Evening Star*, January 28, 1954.

All daylilies are natives of Asia – primarily of China, Japan, and Korea – where they have long been used for medicine and food, and brought to Europe by the 16th century. The botanist Arlow Burdette Stout (1876-1957) arrived at the New York Botanical Garden in 1911 and set about rationalizing the species and hybrids, and cross-breeding new varieties. Today there are more than 80,000 daylily cultivars in just about any color (https://en.wikipedia.org/wiki/Daylily). Ralph Wheeler of Winter Park was a leading hybridizer of the daylily, and his collection dominated the 1945 show ("Mead Garden Daylily Show," *Orlando Evening Star*, April 21, 1945). In 1947, Wheeler was awarded the Herbert Medal for this work ("American Plant Life Society Awards Medal to Ralph Wheeler for his Daylilies," *Orlando Evening Star*, June 25, 1948).

Descriptions of the 1949, 1950 and 1951 shows appear as: "Daylilies Displayed," *Orlando Evening Star*, May 7, 1949; "Show Sunday." *Orlando Evening Star*, May 19, 1950; "Flower Exhibit Draw Crowds at Mead Garden," *Orlando Evening Star*, April 30, 1951; and "Top Hybrids Highlight Daylily Show," *Orlando Evening Star*, May 28, 1951. Wheeler displayed his bluest daylily, Prodigy, at the 1955 show ("Daylily Show," *Orlando Evening Star*, April 29, 1955).

Theo's concern about the plethora of caladium hybrids is from a letter from TLM to Egbert Reasoner, January 6, 1915, RC-TLM. Egbert Norman Reasoner, a world-famed horticulturist, brought more than a thousand varieties of plants, trees, shrubs, vines, and citrus from around the world to this state and is remembered for his introduction of the pink grapefruit. He operated the 200-acre Royal Palm Nurseries in Oneco, FL. The fragrance of the amarcrinum is remarked on in "Mead Caladium Show Will be Held Today," *Orlando Evening Star*, July 21, 1940.

The gift of free blooms at flower shows and the gardenia fragrance in early spring is commented on in "Garden Offers Gardenias," *Orlando Sentinel*, April 30, 1950; "Gardenia Week Celebrated at Mead Botanical Garden," *Orlando Sentinel*, April 26, 1953; and "Mead Giving Free Blooms," *Orlando Sentinel*, April 19, 1953. The *Orlando Evening Star* also reported the first begonia Show, "Garden to Feature Rare Begonias," April 13, 1950, and the first hibiscus Show, "Hibiscus Show," June 21, 1951.

An editorial by Martin Andersen entitled "Mead Garden's Map Prominence," July 5, 1949, praised the new tourist map showing Mead Botanical Garden as the only Orlando attraction of note, but commented that the people of Orlando did not seem to value its beauty. Grover's "Beauty Spots in the Mead Garden," is dated 1941 as a manuscript in RC-TLM, with the inscription "Gift of Miss Frances Grover 4-10-66."

CHAPTER THREE

The approximate number of orchids that Theodore Mead bequeathed to Jack Connery and likely survived until the Garden opened in January 1940 was approximately 1,000 according to Connery's description of the collection in the piece he wrote for the *Orlando Sentinel*, November 14, 1937.

In the early years of the Garden, there was always an experienced orchidist in the greenhouses able to answer visitor's questions about the orchids they were viewing, and provide expert care in their culture. Initially, the role was filled for two years by orchid grower William Anderson, who resigned in 1942 to join the armed forces, passing on the position to William Jess, an experienced gardener from an established estate in the north.

Of the many orchids he hybridized, Mead was perhaps most satisfied with the cross between Cattleya *lueddemanniana* x Cattleya (Laelia) *tenebrosa*, his "Star Boarder," which took 17 years from seed to the first flower, being planted in 1897, first flowered in 1914, and photographed in 1924. This problem-child orchid was reported as blooming at Mead Botanical Garden with sprays of buttercup yellow flowers in "Mead's 'Star Boarder' Orchid Now Blooming with Others in Garden," *Orlando Sentinel*, February 16, 1941, and subsequently, in Edwin Osgood Grover, "The Making of a Botanical Garden," *Parks & Recreation*, (August 1948): 451-456 (RC-TLM). Both of these references appear to be a misidentification – the yellow-flowered orchid referred to was almost certainly Dendrobium *aggregatum* that we know Mead grew.

Cattleya *tenebrosa*, one of the most spectacular of the former Brazilian laelias, was initially described as Laelia *tenebrosa*, a name accepted for nearly a hundred years. Molecular analysis now suggests that the entire group of Brazilian Laelias should be included in the genus Cattleya. This change was published in 2014 in an update to The Orchid Review (Quarterly Supplement to the International Register and Checklist of Orchid Hybrids (Sander's List), January–March 2014 Registrations, distributed with Volume 122, Number 1306, June 2014), and the registered hybridizer of Cattleya *ludbrosa* officially attributed to T. L. Mead.

For more information see Paul Butler, "Theodore L. Mead – Pioneer American Orchid Grower & Hybridizer," *Orchids* 83, no. 1 (2014): 34-37, where there is colored picture of *C. ludbrosa*, Mead's "Star Boarder."

A picture of the rare "blue orchid," Vanda *coerulea*, at Mead Botanical Garden, appears in the *Orlando Sentinel* of January 21, 1943. The original specimen was brought to England from Northern India in 1849 and is still considered rare and difficult to bloom.

The orchid shipment from Baldwin and Company is detailed in a letter from Robert Carr to Edwin Grover, dated November 26, 1939, RC-TLM. In 1940, Carr succeeded Harold Mutispaugh as treasurer of the Garden. The orchids from the Parque Indigena were secured by Charles Williams, a Long Island orchid collector (letter Williams to Grover, February 6, 1940, RC-TLM). The Indigenous Park in Santos, Brazil is a zoological garden reproducing Atlantic Forest vegetation and contains around 3,500 orchids from 120 species, the vast majority being fixed on trees. The orchids were safely received and soon in bloom – "Brazilian Orchids Now in Bloom in Garden," *Orlando Sentinel*, March 14, 1940, as were the Panama Canal Zone orchids ("Thousand Orchid Blooms Now in Mead Garden," *Orlando Sentinel*, April 19, 1942).

The adventures of Tom Sawyer shipping orchids to the USA appears as "Modern Tom Sawyer Hunts Rare Orchid over Jungle Wastes of Venezuela," *Orlando Evening Star*, December 1, 1940. Sawyer used to be one of Ford's most valued salesman in South America before he decided to get into the orchid business. Headquartered in Caracas, Venezuela, he spent six months in the equatorial jungles accompanied by Spanish-speaking natives, collecting orchids for botanists, wholesale florists, and the US Government. Cases contained from 80 to 100 plants were shipped to the USA, some to Mead Botanical Garden where they were divided and potted ("Mead Garden Receives Shipment of Orchids from South America," *Winter Park Herald*, December 28, 1941).

The many varieties of orchids on show were detailed in "Many Varied Orchids are Now Blooming at the Mead Garden," *Orlando Evening Star*, June 16, 1940. It was the Butterfly Orchid (Psychopsis papilio) that supposedly triggered orchid mania – https://en.wikipedia.org/wiki/Psychopsis. Cattleya *J. Palliser* was judged to be the "Most Beautiful Orchid of them All in Bloom at Mead Garden Now," according to the *Orlando Evening Star*, November 10, 1940.

Jack Connery's preference for the Cypripedium orchid comes from "Real 'Men's Orchid' Now Blooming in Mead Garden," *Orlando Evening Star*, November 17, 1940. "Kypria" is Greek for Venus, and podium means "slipper." The two together make Cypripedium, the slipper of the goddess of beauty.

The roots of the American ghost orchid, Dendrophylax lindenii, are so well camouflaged on the tree that the flower seems to float in mid-air, hence its name, "ghost orchid." The plant plays a pivotal role in the non-fiction book *The Orchid Thief* by Susan Orlean. The species is endangered in the wild, and cultivation has proven exceptionally difficult. Today, Florida state law forbids their removal from the wild, but collecting was legal in the early years of the Garden opening, and Connery made several visits to the Everglades to collect this desirable orchid and used them in exchanges with other botanical institutions – see, for example, the article "Mead Garden to be Opened in January," *Orlando Sentinel*, June 11, 1939. The Cigar Orchid and the Florida Butterfly Orchid were similarly widely available in hardwood hammocks and were legally collected in great numbers.

The Million Orchid Project (https://www.fairchildgarden.org/millionorchid), operated by the Fairchild Tropical Botanical Garden, Miami, Florida, seeks to reintroduce millions of young native orchid plants into South Florida using a propagation method co-discovered by Theodore Mead. The description of the orchid incubator display in the greenhouse comes from the piece, "Six-month-old Babies, 'Incubator Orchids,' Seen in Mead Garden," in the *Orlando Sentinel*, November 24, 1940.

Overcrowding in the main greenhouse and the purchase of a second greenhouse principally for cooler-growing orchids is taken from "New Orchid House Now Being Built at Mead to Ease Congestion," *Winter Park Herald*, December 28, 1941, and "Greenhouse Present to Mead Garden," *Orlando Sentinel*, January 11,

1942. The Sherman Adams donation of cymbidiums is mentioned in "Pond for Rare Lilies at Mead Garden Now Under Construction," *Orlando Evening Star*, August 9, 1942. The new Number 3 orchid house appears in "Mead Camellia Show Slated," *Orlando Evening Star*, January 21, 1947.

The American Orchid Society visit is recounted in Nellie Cohen, "Visiting Your Neighbor: The Mead Botanical Garden, Winter Park, Florida," *Orchids* 11, no. 11 (1943): 390-392. Mead's success with hybridizing using Cattleya (now Guarianthe) *bowringiana* as the seed parent, and the naming system he invented is noted in T. L. Mead "Cattleya Bowringiana and its Hybrids," *Orchid Review* 25, no. 289 (1917): 12-13. The Guaranthe genus was segregated from the Cattleya genus in 2003 by Dressler and Higgins based on several structural features of the plants and is supported by recent molecular DNA analyses. The genus comprises the following former Cattleya species: G. *skinneri*, G. *bowringiana*, G. *patinii*, and G. *aurantiaca*. Mead admitted that combining the names of the orchid parents resulted in portmanteau words in a letter from TLM to parents, July 3, 1895 (RC-TLM).

The importance of size in an orchid corsage and its popularity with the wives of US presidents can be found in Ernest Hetherington, "Giants of the Cattleya World," *Orchids* 57, no. 11 (1988): 1204-1215, and A. E. Chadwick and A. A. Chadwick, "First Ladies and Their Cattleyas: *Cattleya* Mamie Eisenhower," *Orchids* 83, no. 5 (2014): 294-297. Jack Connery's greeting with an orchid corsage for Gladys Swarthout is captured in "Diva is Greeted with Orchids," *Orlando Sentinel*, February 16, 1940. Edwin Grover's token of admiration for Lillian Gish appears in "Lillian Gish Gets an Orchid," *Orlando Sentinel*, February 22, 1954, as part of her award of an honorary degree by Rollins College ("Rollins College Awards Honorary Degrees to Six," *Orlando Sentinel*, February 23, 1954).

The description of the Garden in 1948 and its staffing comes from Grover, "The Making of a Botanical Garden," *Parks & Recreation*, (August 1948): 451-456 (RC-TLM). An example of Grover's orchid knowledge is reported in "Grover to Address Sorosis," *Orlando Sentinel*, February 28, 1943. Taylor Briggs' belief that

Grover had a degree in horticulture appears in Gfeller, *The Business of Making Good*, 85.

Record-breaking attendances at the Garden in the late 1940s were due in part to Florida State publicity ("Publicity Boosts Mead Garden," *Orlando Sentinel*, April 6, 1947) and the inclusion of the Garden as a map location site for visitors ("Mead Garden's Map Prominence," *Orlando Evening Star*, July 5, 1949).

CHAPTER FOUR

"The Country Boy's Creed" appears in Gfeller, *The Business of Making Good*, 75. Bacheller's comments on the need to keep in touch with Nature come from "Irving Bacheller Leaves for North with Words for Mead Botanical Garden," *Orlando Sentinel*, May 2, 1939. The allegorical story of a prediction of Chicago's population at the groundbreaking of a canal is taken from "Editorial – City in a Garden," *Orlando Sentinel*, May 22, 1957. The canal opened in April 1848, and the little frontier town of Chicago became the hub of the most important inland trade route in the country, and soon it became one of the fastest growing cities in the world. For the next thirty years, millions of tons of bulk commodities traveled up and down the canal. Grover's plead to preserve natural beauty, helped by public funding, is recorded in "Preserve Central Florida's Natural Beauty," *Orlando Evening Star*, April 1, 1948. In the early years of the 1940s, the Garden received an annual grant of $250 from the City of Winter Park which rose to $500 later in the decade (*Orlando Evening Star*, August 16, 1949).

The Hearst Spanish Monastery story was covered in "Art, Jig-stone Puzzle," *Time*, September 1, 1952, 59, and in "Anybody Want a Monastery?" *Fort Myers News-Press*, February 17, 1946. The Florida State Bureau of Publicity negotiated its relocation to Florida and Winter Park expressed interest ("Winter Park will Dicker for Hearst's $500,000 Monastery," *Orlando Evening Star*, September 14, 1945) and formed an advisory group to acquire it ("Monastery Group Confers on Finances," *Orlando Evening Star*, October 22, 1945). The city approved the Mead Garden location ("City Approves Monastery Site in Winter Park," *Orlando Evening Star*, September 19, 1945) and Hamilton Holt jokingly offered his labor (*Orlando*

Evening Star, October 23, 1945). The building is now a tourist attraction in North Miami (https://en.wikipedia.org/wiki/St._Bernard_de_Clairvaux_Church), with the description of its reassembly in "Art, Jigsaw Puzzle," *Time*, January 11, 1954, 54.

Grover's request for Rollins to take over the Garden is in a letter, Grover to Holt, October 30, 1946, RC-TLM. The planned recreation center was revealed in "City Recreation Center Planned," *Orlando Evening Star*, October 23, 1951, and was strongly promoted by Mayor McCaully; "McCaully Proposal," *Orlando Sentinel*, October 24, 1951, and "Mayor Urges New Park," *Orlando Evening Star*, November 8, 1951. The lukewarm student reception comes from "Winter Park Students Ask Youth Park," *Orlando Evening Star*, April 4, 1952.

Plans to connect Garden Drive with the entrance at Pennsylvania Avenue with a road running through the Garden are contained in "Mead Botanical Garden Reported in Good Shape," *Orlando Evening Star*, January 31, 1952. The additional new parking area promised appears in "Parking Lot Announced – Leedy Tract," *Orlando Sentinel*, March 7, 1952.

"So That the People May Know, by E. O. Grover" was published in the *Winter Park Herald*, June 12, 1952. The conditions under which the city leased the Garden come from "Leased Mead Botanical Garden to Mead Botanical Garden Inc," *Orlando Evening Star*, July 8, 1952. The composition of the new board is revealed in "New Board," *Orlando Sentinel*, December 17, 1952. Connery and Grover expressed the hope that the Garden would become a major botanical center in "Mead Garden Helm Handed to McCaully," *Orlando Evening Star*, December 17, 1952. On December 18, Connery and Grover signed their waivers on the money they had advanced the Garden (*Orlando Evening Star*, December 18, 1952).

The Orange County Board of Education's offer is revealed in "County Offers Only $26,000 for Mead Garden School Site," *Winter Park Herald*, July 30, 1953. The counterproposal appears in "Mead Land School Site," *Orlando Sentinel*, August 6, 1953. Grover's response to the proposal was captured in the *Winter Park Sun*,

August 13, 1953. The *Orlando Sentinel* posed the question "Which would you prefer, to sell part of the land, or lose the Garden entirely?" ("Mead Garden Called Best Asset to Winter Park," *Orlando Sentinel*, August 18, 1953).

The Grover initiative of Rollins taking over the Garden on a 99-year lease is from "Plans Eyed to Lease Mead Garden to Rollins for 99 Year," *Winter Park Sun*, September 24, 1953. The idea of a Rollins College amphitheater in the claypit was also rejected ("Amphitheater Considered for Site in Mead Garden," *Orlando Sentinel*, September 25, 1953). There was a public debate ("Junior High School Site to be Debated," *Orlando Sentinel*, October 11, 1953) at which the majority of the parents present voted for the Mead Garden site ("Parents Back Mead Garden Site for Local School," *Winter Park Sun*, October 15, 1953). The legal status of using land deeded for a botanical garden for a school did not seem to worry the city ("Garden Site for School Gets Backing," *Orlando Sentinel*, October 14, 1953). A slightly improved offer was made for the land ("School Site Question Bolsters Ballot," *Orlando Sentinel*, November 29, 1953) and new directors elected to the Mead Garden Association Board ("Mead Garden Association Elects New Directors," *Winter Park Sun*, November 19, 1953).

"Schools Try Grab of Mead Garden" is how the *Orlando Sentinel* (December 2, 1953) characterized the situation, and this article also contains Grover's issues with using the site for a school and his vision for the Garden's future. The next day's issue of this newspaper featured Grover in a cartoon with Judson and a further rebuff of the school board's proposal ("So the People May Know," *Orlando Sentinel*, December 3, 1953). Voting took place on December 8; ("School Site Ballot Today," *Orlando Sentinel*, December 8, 1953; "Brisk Vote on Mead Issue," *Orlando Evening Star*, December 8, 1953); and the results announced ("Winter Park Votes to Save Mead Garden," *Orlando Sentinel*, December 9, 1953).

The city take-over of Mead Garden was reported in "City Asked to Take Over Mead Botanical Garden," *Orlando Sentinel*, March 3, 1955, and "Winter Park Will Take Over Mead Botanical Operation," *Orlando Sentinel*, March 24, 1955. The city's aim to include a recreation center and community park is recorded in

"Mead Garden May Become City Park," *Winter Park Sun*, March 10, 1955. Legal articles were exchanged in May, ("Instruments Filed in Orange County Records," *Orlando Sentinel*, June 8, 1955) and the setting up of a parks board reported in "City Names Park Board," *Orlando Evening Star*, April 7, 1955. Annexation of the south end of the Garden was concluded late September 1955 ("Annexation Votes," *Orlando Sentinel*, September 20, 1955).

The Connery family's move to Mt. Dora and Jack's efforts growing vegetables at Zellwood were recalled by Edwin Connery (ECC). The land there was so fertile that giant crops could result – examples of beets and carrots grown by Connery were reported (*Orlando Evening Star*, May 5, 1944); and the wonderful taste of Connery's potatoes, and his nickname of "Corncob Jack," was remarked upon when Jack brought a sack of them into the newspaper office (*Orlando Evening Star*, August 18, 1944). The "new center for vegetable production" comes from R. S. Dowdell, "Vegetable Production at Zellwood," Proceedings of the Florida State Horticultural Society, Vol 57, 221-224 (1944).

At Zellwood, Dr. Brown Landone carried on many experiments with the Asian shrub, ramie, demonstrating its unique properties to people and multiple organizations. The plant produces an immensely strong fiber that can be used to weave clothing and was used for wrapping mummies in ancient Egypt. Landone was well known for his scientific research and his work in metaphysics and founded "The Landone Foundation" to publicize his teachings, the major theme of which was to be conscious of new ideas, and to express them in love for the good of all.

Connery's time at the Lake Worth ramie farm was published in "Converted Hemp Harvester is Used for Ramie," *Palm Beach Post*, July 28, 1946. It was the 1940s "wonder crop," with a fiber so strong that Grover challenged Hamilton Holt to break it (Gfeller, *The Business of Making Good*, 83), but the difficulties of mechanically harvesting and processing the fiber proved insurmountable. Connery's time in Cuba, and then Miami, was recalled by Edwin Connery and partially documented in Gfeller, *The Business of Making Good*, 83-84.

The move to DeLand from Miami for the Connery family was triggered by a job opportunity for Jack as landscape engineer at the DeLeon Springs (ECC). The transformation of the site into one of many of Florida's mythological fountains of youth is described in Rick Kilby, "Finding the Fountain of Youth," University Press of Florida, Gainesville, Florida (2013).

The origin of Helen Connery's nickname of "Mimi" comes from the *Orlando Sentinel*, November 13, 1955. Woolworth's extensively advertised their African violet relationship with the Connerys – see for example *Fort Lauderdale News*, November 4, 1955. An example of the perennial call for volunteers appears in "Volunteers Plan Cleanup of Garden," *Orlando Sentinel*, November 5, 1953.

Further information on the Connery's later life comes from personal talks with Edwin Connery, including his recollection of the summer spent tending the orchids, the tendency for orchids to go missing, and his father's deep disappointment with the way the City of Winter Park took over the Garden (ECC). Jack Connery's request to lease Mead Botanical Garden ("Connery Requests Lease," *Orlando Evening Star*, April 28, 1955) was refused, forcing him to consider taking legal action ("Mead Plans to be Set at Meet," *Orlando Evening Star*, May 6, 1955).

CHAPTER FIVE

A brief history of the amphitheater and how "Fashions in the Garden" came about appears in "Winter Park's Amphitheatre is Outstanding Civic Achievement," *Winter Park Sun Herald*, February 18, 1960. The description of the decorations for the first event appears as "New FIG Society Assumes Development of Mead," *Orlando Evening Star*, June 16, 1959. Parking is mentioned in "Orlando Models Named for Fashions in Garden Event," *Orlando Evening Star*, February 15, 1950. Early attendance at the 1950 event was stressed in "Fashions in the Garden to be Gala Spring Event," *Orlando Sentinel*, February 5, 1950, and activities on the day described in "Mead Garden Style Event is Tomorrow," *Orlando Evening Star*, February 24, 1950; "Fashions in the Garden – Style Show of Winter Park Shops,"

Winter Park Topics, February 24, 1950; and "Crowd of a Thousand Expected to Attend Fashions in the Garden," *Orlando Sentinel*, February 19, 1950.

The second annual event was eagerly awaited ("Prizes Will be Awarded at Garden Party," *Orlando Sentinel*, March 17, 1952, and "Don't Forget the Fashions in the Garden," *Orlando Evening Star*, March 20, 1952). The Leo Sunny Trio entertained (*Orlando Sentinel*, March 22, 1952), prizes were awarded, and the Audubon exhibit proved popular ("Top Prizes Awarded by Popular Vote," *Orlando Sentinel*, March 23, 1952).

Reports of the 1953 event can be found as "Wonderland of Flowers is Setting for Preview of Spring," *Orlando Sentinel*, March 8, 1953, and "Garden Will be Show Scene," *Orlando Evening Star*, March 12, 1953. Jonathan Dunn-Rankin, a former Rollins College student, is recognized in https://en.wikipedia.org/wiki/Jonathan_Dunn-Rankin. The 1954 event was covered as "Hospital Donors to Get Garden Pass," *Orlando Evening Star*, February 10, 1954; "Socialites to be Models," *Orlando Evening Star*, February 11, 1954; and "Central Florida Fashion Show," *Orlando Sentinel*, February 11, 1954.

The city's request to turn over the previous year's profits is stated in "Mead Cleanup, Some Still Unhappy," *Orlando Evening Star*, April 8, 1955. A notice of a meeting to discuss the request appears in "Mead Botanical Garden Workers Urged to Meet," *Orlando Sentinel*, April 17, 1955. The newly-built stage comes from "Fashions in Garden to Use Newly-Built Stage," *Orlando Evening Star*, February 21, 1956. The theme was "Easter Parade in Paris" according to the *Orlando Evening Star*, February 16, 1956, and the 1956 profits were put towards permanent seating ("Mead Garden to get Seats," *Winter Park Herald*, May 10, 1956, WPPL.

The "Roman Holiday" of 1957 is covered in "Fashions in the Garden Take a Roman Holiday," *Orlando Sentinel*, February 24, 1957. The surprise opening is described in "Garden Fashion Show Held Best Yet," *Orlando Evening Star*, March 5, 1957. In 1958, Edith Tadd Little ("Winter Park Town Crier," *Orlando Sentinel*, February 5, 1959) assumed responsibility for set design. The event is documented in

"Queen for a Day Reigns at Fashions in Garden," *Orlando Evening Star*, March 27, 1958. The profits went towards lighting for the amphitheater (*Orlando Evening Star*, April 17, 1958) with the old lampposts from Park Avenue donated by the Florida Power Corporation ("New FIG Society Assumes Development of Mead," *Orlando Evening Star*, June 16, 1959).

The organization of the 1959 show comes from "FIG Plans Laid by Garden Clubbers," *Orlando Sentinel*, March 4, 1959, with a Gone-with-the-Wind experience described in "Could Rumor be True?" *Orlando Sentinel*, March 20, 1959. Agreement to fund the amphitheater dressing rooms can be found at "Mead Gets Dressing Rooms," *Orlando Evening Star*, June 4, 1959, and "Dressing Rooms Financed," *Orlando Evening Star*, July 8, 1959. The builder was Allen Trovillion ("Winter Park's Amphitheater is Outstanding Civic Achievement," *Winter Park Sun Herald*, February 18, 1960, RC-TLM). News of the amphitheater dedication can be found at "Quality Show to Dedicate Mead Amphitheater," *Orlando Sentinel*, September 13, 1959, and at "FIG Amphitheater Dedication Tomorrow," *Orlando Sentinel*, September 30, 1959.

The description of "Aloha to Hawaii" in 1960 comes from the following sources: "Aloha to Hawaii is Fashion Theme for Garden Show," *Orlando Evening Star*, January 28, 1960; "Beautiful Prizes Add Excitement for Fashions in Garden Festival," *Orlando Sentinel*, February 14, 1960; "Native Islander Shows Variety of Hawaiian Fashions," *Orlando Evening Star*, March 23, 1960; "Hawaiian Designs Spark Fashions in the Garden," *Orlando Sentinel*, April 3, 1960; "Ninth Fashions in Garden Rated Best Ever Presented," *Orlando Evening Star*, April 12, 1960; and "Among Outstanding Styles at FIG," *Orlando Evening Star*, April 13, 1960.

For 1961, the show was covered as "Fashions in Garden to Feature Home Town, USA," *Orlando Sentinel*, March 19. For 1962, April 7 was the chosen date where a "Galaxy of Styles Coming Saturday to Mead Garden," announced the *Orlando Sentinel* on April 1, 1962, and "Stage Set Tomorrow for Fashions in Garden," on April 6. Rain caused a postponement for a week ("Spring Showing of Fashions in the Garden," *Orlando Evening Star*, April 11, 1962) and the day before the rain date of April 14, "Green Thumbers Looking Skyward," noted the *Orlando Sentinel*. The "Caribbean Cruise" of 1963 is described in "Fashions in Garden Planned," *Orlando Sentinel*, January 13; "…On a Caribbean Cruise," *Orlando Sentinel*, March 17; "Fashions in the Garden Scores Triumph," *Orlando Evening Star*, April 1; and "Fashion Show, Club Events Bring Out Area Residents," *Orlando Evening Star*, April 2.

Talk of a fall event appears in "Fall Fashions May Grace Gardens," *Orlando Evening Star*, June 28, 1963. Aspects of the family day are described in "FIGs to Honor Great Gardener," *Orlando Evening Star*, February 6, 1964; "Saturday is Family Day in Mead Botanical Garden," *Orlando Sentinel*, May 3, 1964; and "Families in Garden Benefit Set Tomorrow," *Orlando Evening Star*, May 8, 1964. The fashion show returned in 1975 ("Fashions in Garden Coming Back," *Orlando Sentinel*, March 2, 1975) and subsequent years. In 1979 it was reported as "Garden Club Fashion Show Draws Sun, Good Turnout," *Orlando Sentinel*, April 1, 1979. Notice of the last one in 1987 can be found in RC-WPGC (A ticket for the event on April 1, 1987). The Orlando Fashion Mall opened in 1973 ("Opening of Fashion Mall," *Orlando Sentinel*, July 29, 1973).

CHAPTER SIX

Residents complaining about snakes comes from "Snake Pit Sets Off Blast at Councilmen," *Orlando Evening Star*, March 18, 1955, and the news of residents taking matters into their own hands appears in "Snake Pit Cleanup Promised by City," *Orlando Evening Star*, March 22, 1955. Extension of the cleanup and the final city conclusion is documented in "Rats Swarm Out from Mead Garden Snake Pit," *Orlando Evening Star*, April 1, 1955, and "Mead Cleanup Ended, Still Some Unhappy," *Orlando Evening Star*, April 8, 1955.

Establishing a tree and plant nursery in the Garden is reported in "Commission Sets Up Tree Nursery for City," *Orlando Evening Star*, May 15, 1957, and "Nursery Problems Outlined," *Orlando Evening Star*, September 9, 1964. Moving it to another location is in "Oops! There Goes, City to Move Rubber Plants," *Orlando Evening Star*, February 23, 1968.

McCaully's and the Winter Park Garden Club's concerns are stated in "Mead Plans to be Set at Meet," *Orlando Evening Star*, May 6, 1955; waiving the entrance fee and developing the Garden in "Club Postpones Acceptance of Mead Garden Lodge," *Orlando Evening Star*, May 10, 1955; and "Mead Garden Arrangement Set," *Orlando Evening Star*, May 12, 1955. The city efforts to renovate and improve the Garden are listed at "Mead Garden Renovated," *Orlando Evening Star*, July 14, 1955; "Problem of Cleaning Lagoon Lessens," *Orlando Evening Star*, May 14, 1955; and "Parks Board Maps Ambitious Undertaking," *Orlando Evening Star*, December 1, 1956.

The request by the Florida Audubon Society comes from "Board Approves Plans for Audubon Nature Study Center in Garden," *Orlando Evening Star*, May 3, 1957, and "Audubon Society Building," *Orlando Evening Star*, July 5, 1957. The city's refusal is recorded in the *Orlando Evening Star*, October 3, 1957. Information on the headquarters of the FFGC building comes from "Winter Park Offers FFGC Building Site," *Orlando Sentinel*, February 2, 1959; "Groundbreaking Set Jan 6," *Orlando Sentinel*, December 1, 1959; "Gold Spade Starts New Garden Bldg.," *Orlando Evening Star*, January 7, 1960; and "New $70,000 FFGC Bldg. Opens," *Orlando Evening Star*, September 1, 1960.

Arguments concerning a new elementary school location, and the alternative Banks site, occupied many newspaper articles in 1959; "New Elementary School Eyed," *Orlando Evening Star*, July 9, 1959; "Board Opposes School in Mead, Clay Pit Deal Due Fight," *Orlando Evening Star*, July 10, 1959; "Women to Protest Mead School," *Orlando Sentinel*, July 12, 1959; "Banks Site for School Cheered," *Orlando Evening Star*, July 16, 1959; "City Leaders Praise Garden Site Selection," *Orlando Sentinel*, July 17, 1959; "School Site Parley Awaits Site Appraisal," *Orlando Evening Star*, July 23, 1959; and "Co-op Apartments Planned," *Orlando Evening Star*, January 5, 1961.

Grover's outburst against the city for refuse dumping and handling appears in "Dr. Grover Deplores Mead Destruction," *Orlando Evening Star*, July 15, 1959, and "Mead Garden Has Suffered, Grover Says," *Orlando Sentinel*, March 13,

1961. The refuse transfer facility was the last straw and violently opposed by residents ("Contract Awarded for Refuse Handling," *Orlando Sentinel*, March 5, 1961; "Transfer Location to be Decided," *Orlando Evening Star*, April 13, 1961; and "Mead Trash Transfer Opposed," *Orlando Sentinel*, April 16, 1961).

Grover's visits to the Garden with Frances and his yardman are recollected by Taylor Briggs and published in Gfeller, *The Business of Making Good*, pp. 84-85. The *Sentinel-Star* orchid initiative is described in "New Mothers Get Orchids," *Orlando Sentinel*, March 16, 1962; "Sentinel-Star Retail Orchid Center Called 'Year-Round Show' of Blooms," *Orlando Sentinel*, November 16, 1963; and "How Orchid Center Got into Business," *Orlando Sentinel*, November 24, 1963. The dedication of the Grover trail marker is in "Garden Trail Dedicated to Dr. Grover," *Orlando Sentinel*, June 25, 1961, and his birthday celebrations described in "Mead Garden Founder is 93," *Orlando Evening Star*, June 5, 1963.

Archibald Granville Bush made an $800,000 gift to Rollins College in 1965, making the Bush Science Center possible. In 1909, he joined the fledgling 3M company and started buying 3M stock, and by 1959 he owned about $80 million's worth. In 1949 he purchased a home in Winter Park and served on the board of Rollins and Winter Park Memorial Hospital, donating generously to both. Grover's letter to Bush is dated December 28, 1963, and resides in RC-TLM. The report of Grover's death can be found at "Dr. Grover, Leader at Rollins, Dies," *Orlando Evening Star*, November 9, 1965.

The creation, planting, and incorporation of the figure of St. Francis at the Shippen Retreat is covered in "Shippen Retreat Rites Slated for Saturday," *Orlando Evening Star*, June 16, 1959; "Shippen 'Quiet Retreat' Dedicated in Simple Mead Garden Rites," *Orlando Evening Star*, June 23, 1959; and "At Shippen Retreat," *Orlando Sentinel*, November 8, 1959.

The meeting to discuss long-term improvements in Winter Park is reported as "$2.5 Million in Bonds Suggested by Citizens," *Orlando Evening Star*, September 16, 1959. At the start of 1960, the Garden became free ("City's Mead Garden

Now Free," *Orlando Sentinel*, January 24, 1960) and attracted record attendance ("Visitors Set Record High at City Park," *Orlando Evening Star*, January 27, 1960).

Information on Mac McConnell comes from a number of sources. He was a prolific letter-writer to the *Orlando Sentinel* over the period 1952-1968, displaying his hatred of communism ('red vipers,' letter, *Orlando Sentinel*, January 12, 1964) but also his appreciation of the friendship offered by Edwin Grover (letter, *Orlando Sentinel*, November 18, 1965). The replacement of his wooden shed by a concrete building at the entrance to the Garden and its refurbishment is covered in "Exchanging the Old for the New," *Orlando Evening Star*, April 1, 1957; "Furniture Request at Garden," *Orlando Evening Star*, March 29, 1957; and "Sewer Extension Job Leaves Street Rough," *Orlando Evening Star*, May 17, 1957. His role in the Garden is mentioned in "Mead's 'Mac' is Rare Man," *Orlando Evening Star*, September 29, 1959, and "Orchid House Life Teaches Patience," *Orlando Evening Star*, March 30, 1967. William Ferrigno took charge of the orchids when Mac retired ("Speaking of Gardeners," *Orlando Sentinel*, March 20, 1969). McConnell's obituary appeared in the *Orlando Sentinel*, December 20, 1987.

Mayor Pflug's request for garden volunteers is noted in "Volunteers Sought on Mead Garden Job," *Orlando Sentinel*, May 18, 1960. Claypit beautification by the city appears in "Garden Clay Pits Get Beauty Treatment," *Orlando Evening Star*, August 9, 1960. Bev Brown's efforts to restore the orchid houses are reported in "Orchid Houses to be Expanded," *Orlando Sentinel*, September 14, 1963; "New Hours for Mead Tours Told," *Orlando Evening Star*, November 1, 1963; and "Mead to Revamp Orchid Houses," *Orlando Evening Star*, September 13, 1963.

Bricks taken up from Winter Park's streets and piled in the Garden were commented on in the *Orlando Evening Star*, November 19, 1963. A picture of the final brick being laid to complete the driveway appears in "80,000th Brick Laid," *Orlando Evening Star*, March 31, 1964. The garden transformation into a community park was set out in "Garden Now Allows Cookouts," *Orlando Evening Star*, March 11, 1964.

Statements of vandalism in the Garden include "Police Check 2 Cases," *Orlando Evening Star*, December 31, 1957; "Police Kept on Jump by Varied Violations," *Orlando Evening Star*, May 20, 1958; "Vandals Hit Mead Garden, Extensive Damage Reported," *Orlando Evening Star*, May 31, 1966; "Vandals Hit Mead Garden," *Orlando Evening Star*, January 4, 1967; "Vandals at Mead 2nd Time in Week," *Orlando Evening Star*, October 2, 1968; "Winter Park Vandalism Problem Grows," *Orlando Sentinel*, January 12, 1969; "Mystery Not Solved but Statue Returned," *Orlando Sentinel*, January 9, 1969; and "Mead Garden St. Francis Again 'Kidnap' Victim," *Orlando Sentinel*, January 17, 1969.

Reports of classical music concerts in the amphitheater appear in newspapers of the time, for example, "Music Fills Sunday Skies for Mead Theater Crowd," *Orlando Evening Star*, February 16, 1960, and "Symphony to Give Concert on Sunday," *Orlando Evening Star*, April 13, 1967.

The master plan of 1967 developed over the year, started with "Development Plan to be Aired on Mead Garden," *Orlando Sentinel*, February 19, 1967; "City Won't Become Road Hub," *Orlando Evening Star*, Aril 19, 1967; and culminated with "Douglass Lauds Planning for Future of Winter Park – Five-Year Budget Program," *Orlando Evening Star*, June 11, 1968. A description of the elements of the 1967 master plan can be found in an undated manuscript in RC-TLM.

In the early years, the N-S road to the west of the Garden was named "Maitland Avenue." At a regular meeting of the Winter Park commission on April 8, 1975, a resolution was passed renaming the road Denning Avenue in honor of Girard Denning, a native of Winter Park who during his lifetime served as Mayor for a period and as Postmaster for many years.

CHAPTER SEVEN

Taking small steps in garden improvements seemed to make sense to Helen Dunn-Rankin ("Little Projects Aid Mead Garden," *Orlando Evening Star*, January 19, 1957). The five-year master plan fizzled out according to "Mead Garden Expand Plans Gather Dust," *Orlando Sentinel*, May 16, 1971. Abandonment of the Dinky

Line is covered in "Dinky Line Abandonment Announced by SCL Railroad," *Orlando Sentinel*, November 26, 1967, and "Winter Park Soon to Get Dinky Line Right-of-Way," *Orlando Sentinel*, November 23, 1967. The Sevilla subdivision development comes from "Rezoning Hearing to be Resumed," *Orlando Evening Star*, December 29, 1967; "High-Cost Housing Proposed," *Orlando Evening Star*, December 18, 1969; and "Sevilla Debut Due Today," *Orlando Sentinel*, November 15, 1970.

Alternative uses of the Dinky Line were put forward and rejected in "Dinky Line Plan Travels Rocky Route," *Orlando Evening Star*, January 26, 1967; "Dinky Line Park Tour Eyed," *Orlando Sentinel*, December 31, 1966; "Trio Scorns Dinky Line's Use as Park," *Orlando Sentinel*, January 24, 1967; and "Dinky Line Blast Called Premature," *Orlando Sentinel*, January 26, 1967.

References related to the new Winter Park Garden Club building are "Mead Garden Entry Arrangement Set," *Orlando Evening Star*, May 12, 1955; "Site of Garden Center Addition," *Orlando Evening Star*, June 21, 1961; "Garden Building $$ Drive Gets Under Way," *Orlando Sentinel*, November 24, 1969; "Mead Garden Expand Plans Gather Dust," *Orlando Sentinel*, May 16, 1971; "Survey for New Garden Club," *Orlando Sentinel*, November 12, 1972; and "Garden Club Building Blooms," *Orlando Evening Star*, January 10, 1973.

Lee Simpson helped look after orchids in the greenhouse in 1969 ("Places to Find Peace…Where Serenity Abounds," *Orlando Sentinel*, December 22, 1969). The Blanchard call for more small parks comes from "Winter Park Needs More Parks," *Orlando Evening Star*, September 12, 1969. His rejected request for more park workers and his subsequent resignation appear in "Request for Additional Workers Questioned," *Orlando Evening Star*, August 30, 1972, and "Jay Blanchard Quits, Accepts County Post," *Orlando Sentinel*, February 14, 1973.

The consequences of the decision to combine the parks and public works activities are found in "Buck Accepts Top Job," *Orlando Sentinel*, June 29, 1974; "Job Offers, New Interest Spur Changes," *Orlando Sentinel*, June 21, 1974; "Units Should be Merged," *Orlando Sentinel*, June 21, 1974; and "Parks, Recreation

Reunion Recommended," *Orlando Sentinel*, July 14, 1974. The city cutbacks lead to concerns about safety ("City Budget Lacks Safety Unit Cuts," *Orlando Sentinel*, September 30, 1976) and only one employee maintaining the Garden ("Mead Garden – Garden Club, Mayor Differ on Whether Winter Park Showcase is Neglected Area," *Orlando Sentinel*, May 11, 1977).

The last pictures found of the original greenhouses date from 1973 and 1974. Greenhouse #3 by the bank of the south lily pond appears in the *Orlando Sentinel*, February 16, 1973; a picture of the dilapidated interior entitled "Greenhouse Enjoyed by Visitor" can be found in the *Orlando Sentinel*, April 11, 1973; and "Orchid Queen" shows Betty Mobley in the *Orlando Sentinel*, April 17, 1974. William Ferrigno's retirement is celebrated in "Retiring Again," *Orlando Evening Star*, February 2, 1972. Movement of the orchids off-limits, and the comments by John Holland, Charles Sheppard, and James Driver, come from "Garden Club, Mayor Differ on Whether Winter Park Showcase is Neglected Area," *Orlando Sentinel*, May 11, 1977.

The YCC activities are recorded in "Youth Conservation Corps Landscaping School, Gardens," *Orlando Sentinel*, July 25, 1976, and "Conservation Teens Earn, Work, Learn," *Orlando Sentinel*, August 13, 1976. The Winter Park Chamber of Commerce report confirmed the neglect ("Improvement at Mead Garden," *Orlando Sentinel*, July 6, 1977) but noted an improvement in tidiness following the YCC efforts ("Youths Improve Mead Garden," *Orlando Sentinel*, July 3, 1977).

Three new YCC projects were suggested for 1978 ("Traffic Rough on Mead Garden Trees," *Winter Park Sun Herald*, November 17, 1977) of which two were approved; a new greenhouse ("Greenhouse Approved," *Orlando Sentinel*, November 13, 1977) and a new boardwalk ("Youths Bridge Gap," *Orlando Sentinel*, August 2, 1978; "Garden Locked Gate Oversight," *Orlando Sentinel*, September 6, 1978). In 1980, volunteer help switched to the Scouts ("Mead Garden Target of Scout Cleanup Project," *Orlando Sentinel*, September 3, 1980).

The "severe identity crisis," description of neglect, and plea for community help appears in "Help Sought in Campaign to Revive Mead Garden," *Orlando Sentinel*,

May 7, 1982. The story of the sale of what remained of Mead's orchids comes from "Orchid Sale to Raise Money for Mead Garden," *Orlando Sentinel*, May 12, 1982, and "Looking for Help for Mead Garden," *Orlando Sentinel*, May 26, 1982. $500 for one of Mead's orchids is referenced in Butler, *Orchids and Butterflies*, 270. Shamed by the criticism of neglect, city crews started tidying up ("Winter Park Puts City Crews to Work on Mead Garden," *Orlando Sentinel*, November 26, 1982). The suggestion of improving the Winter Park entrance to the Garden was made in "Group Using Award to Build Support for Mead Garden," *Orlando Sentinel*, June 22, 1983. In 1985, the city's effort is recorded as "Mead Garden Gets a Facelift," *Orlando Sentinel*, May 17, 1985; the actions of volunteers as "Groups Plan to Spruce up Mead." *Orlando Sentinel*, November 8, 1985. The Winter Park High School's Future Farmers of America Club renovated the only greenhouse left at the Garden in "Orchid House Idea Blossoms into Big Student Project," *Orlando Sentinel*, February 16, 1986. Rollins students and residents cleaned up and repaired the boardwalk ("Volunteers Restoring Mead Garden," *Orlando Sentinel*, April 29, 1987.

Trovillion's land grab attempt comes from "Deed Limits Winter Park's Options in Mead Garden," *Orlando Sentinel*, December 13, 1987. The Winter Park Historical Society's proposed move to the Garden is assembled from "Historical Society Considers Mead Garden for Museum Site," *Orlando Sentinel*, March 7, 1991; "Looking for Long-Lost Relatives," *Orlando Sentinel*, April 7, 1991; and "History Group Ponders Mead," *Orlando Sentinel*, July 18, 1991. Fundraising with an old map of Winter Park and the new location for the museum come from "Old Map May Lead to New Home for Historical Society," *Orlando Sentinel*, April 4, 1993.

The Lake Lillian alligators and their eventual capture graced the newspapers in "Mead Garden Opens Season," *Orlando Evening Star*, November 11, 1946; "Mead Garden Renovated," *Orlando Evening Star*, July 14, 1955; "Mead Garden Menace Held," *Orlando Evening Star*, August 28, 1964; "Man vs. Gator at Mead Garden," *Winter Park Observer*, August 13, 1992; and "8-Foot Alligator Gets a Transfer," *Orlando Sentinel*, July 29, 1992.

The boardwalk repairs were put on hold in "City Park to Get Cleanup," *Orlando Sentinel*, November 13, 1987. By then it was time for another master plan as detailed in "City Plans Park for all Seasons," *Orlando Sentinel*, December 23, 1988; "Long-range Upgrading of Mead Garden Set," *Winter Park Outlook*, January 26, 1989; "City Needs Lots of Help to Polish Mead Garden," *Orlando Sentinel*, May 5, 1989; and "Volunteers Want Neglected Mead Garden to Blossom Again," *Orlando Sentinel*, January 7, 1990.

CHAPTER EIGHT

"Teen Volunteers Work Hard to Clear Mead Garden Creek," *Orlando Sentinel*, July 16, 1989, describes Operation Comeback and the preparation for the boardwalk. Donations were received from the Winter Park and International Lions Club for the Braille trail ("Mead Garden Lets Blind Touch Nature," *Orlando Sentinel*, March 26, 1992), and a new logo for the Garden created ("Mead Garden Receives $1,000 Check and Unveils New Logo," *Winter Park Observer*, March 26, 1992). Work began in earnest in 1993 to construct the boardwalk loop which was mapped and sensory stations identified ("Mead Trail Will Give Senses a Workout," *Orlando Sentinel*, February 28, 1993; "Get a Feel for Nature on Gardens Boardwalk," *Orlando Sentinel*, September 30, 1993; and "Mead Garden Boardwalk Dedicated," *Winter Park Observer*, October 7, 1993). In 2002, additional improvements were added ("New Multi-sense Trail in Works at Mead Garden," *City of Winter Park Update*, June/July 2002, accessed April 9, 2019 at https://cityofwinterpark.org/docs/media/newsletters/winter-park-update/June_02.pdf).

The fast-growing invasive skunk-vine (also known as stinkvine) probably started to infest the wetlands of the Garden in the late 1980s, growing steadily before expanding rapidly in 1992 ("Fast-growing Vine at Root of Park's Troubles," *Orlando Sentinel*, October 10, 1992). The botanical name of the vine is *Paederia foetida*, and once established vegetative spread from creeping stems can be rapid and extensive, with stem fragments capable of rooting and producing new vines. The extent of the infestation remained unknown to the city, and it required concerned members of the Winter Park Garden Club to bring it to their attention (Letter, Roberts and Leonard to Mayor Johnson, September 30, 1992, RC-WPGC). A second category one invasive was also present – the air potato vine – *Dioscorea bulbifera*.

In 1995, another call for volunteers to "pull" vines took place ("Volunteers Work Wonders at Mead Garden," *Winter Park Observer*, November 9, 1995), and in 1997 it was the Mennonites' turn ("Entire City is Mennonites' Convention Center," *Orlando Sentinel*, July 28, 1997, and "1,000 Mennonites Brave Heat, Skunk Vines in Winter Park," *Orlando Sentinel*, July 31, 1997). The city switch to herbicide comes from "New Offensive Targets Skunk Vine," *Orlando Sentinel*, December 27, 1998. Herbicide treatment and retreatment was necessary for the next seven years, and the vine canopy resulted in many more trees in the wetlands being blown down when hurricane Charlie hit the Garden in 2004 (Tim Egan, private communication, 2018).

The efforts that went into the 2007 master plan being accepted by the community appears in "Mead Garden Master Plan Public Meeting," *Winter-Park-Maitland Observer*, September 19, 2006. The priority of the different elements of the plan comes from the City of Winter Park, Parks and Recreation Department Strategic Plan 2009–2014, https://cityofwinterpark.org/departments/parks-recreation/administration/publications.

Randy Knight graduated from Texas A&M University with a degree in floriculture and horticulture. He owned Poole and Fuller, a nursery and then landscaping business in Winter Park for 25 years, and had for a time the contract to keep Park Avenue's landscape beautiful. For years he was the horticultural advisor to the Albin Polasek Museum & Sculpture Gardens, and once he retired this became his full-time occupation. In 2012, he started the horticultural transformation of Mead Botanical Garden.

The likening of the Garden to an unwanted orphan was made in "Citizens Voice Their Needs," *Orlando Evening Star*, September 16, 1959. "Hope springs eternal" is a phrase from the Alexander Pope poem *An Essay on Man* and suggests that it is a characteristic of human nature always to find fresh cause for optimism.

Picture Credits

T= top, M = middle, B = bottom, L = left, R = right, TR = top right, BR = bottom right, BL = bottom left.

Front and back cover: Postcard images of Mead Botanical Garden, circa 1940s. Photographs by T. P. Robinson, Orlando, Florida and published by Hartman Litho Sales, Box 697, Largo, Florida.

Frontispiece: Photographs of Jack Connery and Edwin Grover, courtesy of the Department of College Archives and Special Collections, Olin Library, Rollins College, Winter Park, Florida.

1.1T, 1.1BL, 1.2, 1.12B, 1.13, 2.1B, 2.4BR, 4.5T, 4.5BR: Courtesy of the Edwin Connery and family collection.

1.1BR, 1.5, 1.7, 1.8, 1.9, 1.11, 1.12T, 1.14, 2.4T, 2.6, 2.9L, 2.13, 3.1B, 3.2, 3.6, 3.10, 3.11T, 4.1, 5.2L 5.4B, 5.5, 6.1, 6.2, 7.1, 7.3, 7.5T, 8.3TL, 8.3R: Used with permission, courtesy of the Department of College Archives and Special Collections, Olin Library, Rollins College, Winter Park, Florida.

1.3: From Paul Douglass and Alice McMahon, "Recreation Plan for the City of Winter Park," Center for Practical Politics, Rollins College, 1960.

1.4: From Rick Singh, Orange County Property Appraiser, webmap at www.ocpafl.org.

1.6, 7.7, 8.1, 8.2TL: Used with permission, Winter Park History and Archive Collection, Winter Park Public Library, Winter Park, Florida.

1.10L: From US Library of Congress, Prints & Photographs Division, 1917.

1.10M: Used with permission, courtesy of the First Unitarian Church of Orlando, 1954.

1.10R, 2.4BL, 2.11L: Used with permission, courtesy of the Orange County Regional History Center.

1.15: From the Orlando Evening Star, November 2, 1939.

1.16, 2.3L, 8.4: Undocumented newspaper clippings, Department of College Archives and Special Collections, Olin Library, Rollins College, Winter Park, Florida

2.1L: Wikimedia Commons File:Rhododendron austrinum 1zz.jpg.

2.1R: From www.bartramsgarden.org.

2.2: Postcard image, Curt Teich & Co., Chicago, 1940.

2.3R, 2.5, 3.1T: Postcard image, Orange News Co., Orlando, Fl., circa 1940s.

2.7: From City of Winter Park, Engineering Department, "Topographic Survey of Mead Garden," April 1965.

2.8L, 2.8R, 2.10R, 2.12, 3.5M, 3.8L, 5.3, 5.6, 7.4, 8.2TR, 8.6R, 8.9: From author's collection.

2.8M: From the University of Connecticut, Biodiversity Education & Research Greenhouse, Florawww.eeb.uconn.edu/201600167.html.

2.9R: From the Miami Daily News, Sunday, May 11, 1941, photograph by Charles O'Rork.

2.10L: Wikimedia Commons File:Kamelien-Königsbrück-Weißblüte.jpg.

2.11TR, 2.11BR: The American Daylily Society, htttps://daylilies.org/awards/stout-silver-medal, photos by Oliver Billingslea.

3.3, 3.4, 3.7R: From R. Warner and B. S. Williams The Orchid Album, Published by B. S. Williams, London in 11 volumes, 1882-1897.

3.5L: Wikimedia Commons File:Dendrophylax_lindenii_(16979925623).jpg.

3.5R: Wikimedia Commons File:Encyclia_tampense_South_Miami_USA.jpg.

3.7L: From Nellie Cohen, "Visiting your neighbor: The Mead Botanical Garden, Winter Park, Florida" American Orchid Society Bulletin, vol 11, No. 11, pp 390-392, 1943.

3.8R: Copyright Greg Allikas, used with permission, www.orchidworks.com.

3.9L: From the Winston-Salem Journal, May 8, 2014.

3.9R: From the Orlando Sentinel, February 16, 1940, photograph by Charles O'Rork.

3.11BL: From the Orlando Post, May 2, 1948.

3.11BR: From the Orlando Sentinel, December 2, 1953, Sentinel Foto.

4.2T: From www.spanishmonastery.com/museum-affairs.

4.2B: Wikimedia Commons File:Monastery_of_St_Bernard_de_Clairvaux_9.jpg.

4.3: From the Orlando Evening Star, December 18, 1952, Sentinel-Star photo.

4.4: From the Orlando Sentinel, December 3, 1953.

4.5BL: From the Fort Lauderdale News, November 4, 1964.

5.1: From the Orlando Sentinel, March 19, 1952.

5.2R: From the Orlando Evening Star, June 16, 1959.

5.4T: From the Orlando Sentinel, March 15, 1959.

6.3: From the Winter Park Sun Herald, June 29, 1961.

6.4: From the Orlando Evening Star, June 5, 1963.

6.5: From the Orlando Sentinel, November 8, 1959.

6.6: From the Orlando Evening Star, April 1, 1957, photographs by Ward.

6.7: From the Orlando Sentinel, May 16, 1971.

7.2L: From the Orlando Evening Star, May 5, 1971, photograph by Don Meade.

7.2R: From the Orlando Sentinel. April 17, 1974, photograph by Bob Frey.

7.5M: From the Orlando Sentinel, August 2, 1978.

7.5B: From the Orlando Sentinel, May 7, 1982, photograph by George Skene.

7.6L: From the Orlando Evening Star, August 28, 1964, photograph by Frank Godwin.

7.6R: From the Orlando Sentinel, July 29, 1992, photograph by George Skene.

8.2L: From the Orlando Sentinel, February 28, 1993, photograph by George Skene.

8.3BL: From the Orlando Sentinel, December 27, 1998, photograph by John Raoux.

8.5: Tim Egan, City of Winter Park, private communication.

8.6L: Photograph by Chris Evans, University of Illinois, Bugwood.org.

8.6M: From https://wildlifeofhawaii.com/flowers/1965/dioscorea-bulbifera-air-yam.

8.7: Map redrawn from single page manuscript, "Mead Garden Nature

Preserve," courtesy of the Department of College Archives and Special Collections, Olin Library, Rollins College, Winter Park, Florida.

8.8: From https://cityofwinterpark.org/docs/government/boards/agendas/PRAB-agd-2014-11-19.pdf.

Acknowledgment image: Author's collection.

Also by Paul Butler

Email contact for book purchase inquiries is casajard@gmail.com.

www.ingramcontent.com/pod-product-compliance
Lightning Source LLC
Chambersburg PA
CBHW061141010526
44118CB00026B/2834